Early Estimation of Project Determinants

Predictions through Establishing the Basis of New Building Projects in Germany

von

Dr.-Ing. Onur Dursun

Oldenbourg Verlag München

Editor: Dr. Stefan Giesen
Production editor: Tina Bonertz
Cover design: hauser lacour

Dissertation, Universität Stuttgart (D 93), 2013

Library of Congress Cataloging-in-Publication Data
A CIP catalog record for this book has been applied for at the Library of Congress.

Bibliographic information published by the German National Library
The German National Library lists this publication in the Deutsche Nationalbibliografie; detailed bibliographic data are available on the Internet at http://dnb.dnb.de.

Rosenheimer Straße 143, 81671 München, Germany
www.degruyter.com/oldenbourg
Part of De Gruyter

Printed in Germany

This paper is resistant to aging (DIN/ISO 9706).

ISBN 978-3-11-034638-1
eISBN 978-3-11-034787-6

Early Estimation of Project Determinants

Predictions through Establishing the Basis of New Building Projects in Germany

Faculty of Architecture and Urban Planning, University of Stuttgart
A thesis submitted in fulfillment of the requirement for the award of the Degree of Doctor of Engineering (Dr. Eng.)

by
Onur Dursun
born in Izmir, Turkey

Supervisor: Prof. Dr. Christian Stoy
Co-Supervisor: Prof. Dr. Arnold Tautschnig
Co-Supervisor: Prof. Dr. Michael Oberguggenberger

Date of Defense: 22.August.2013

Institute for Construction Economics, University of Stuttgart
2013

Contents

List of Abbreviations

AFS	Average floor size
ANNs	Artificial neural networks
ANOVA	Analysis of Variance
APE	Absolute percentage error
ASH	Average storey height
BKI	Cost Information Centre of German Chamber of Architects GmbH
BTC	Bromilow's time-cost model
CBR	Case-based reasoning
CG	Cost group according to hierarchical structure of DIN 276-1
CT	Client type
DPA	Developed plot area
FAC	Facility type
GBV	Gross building volume
GC	Ground conditions
GEFA	Gross external floor area
GRN	General regression neural net
HOAI	Regulation on fees for architects and engineers
iid	independently and identically distributed
k300	cost of structure - construction works
k300+k400	cost of structure
k310	cost of excavation
k320	cost of foundations
k330	cost of external walls
k340	cost of internal walls
k350	cost of floors and ceilings
k360	cost of roofs
k370	cost of structural fitments
k390	cost of other construction related activities
k400	cost of structure - services
k410	cost of sewerage, water and gas systems
k420	cost of heat supply systems
k430	cost of air treatment systms

k440	cost of power installation systems
k450	cost of telecommunications and other communication systems
k460	cost of transport systems
k470	cost of function related equipment and fitments
k480	cost of building automation
k490	cost of other services related work
LI	BKI location index
LRM	Linear regression models
LSE	Least square estimate
m310	Excavation volume (cubic meters)
m320	Foundation area (square meters)
m330	External walls area (square meters)
m340	Internal walls area (square meters)
m350	Floors and ceilings area (square meters)
m360	Roof area (square meters)
MAD	Mean absolute deviation
MAER	Mean absolute error rate
MAPE	Mean absolute percentage error
MC	Market conditions
MLF	Multi layer feedforward neural network
MLP	Multi layer perceptrons
MSE	Mean square error
NoS	Number of storeys
NoSag	Number of storeys above ground
NoSbg	Number of storey below ground
PA	Plot area
RMSE	Root mean squared error
RUQ	Reference unit quantities
SA	Site access
SMEs	Small and Medium Size Enterprises
TOP	Topographical conditions
€	Currency unit of European Union

List of Figures

List of Tables

Abstract

Construction cost and duration are two strategic determinants of building projects along with quality. Objective estimations of these determinants are crucial through establishing the basis of a project, hence they serve as foundation for budgeting, planning, executing, monitoring and even for any litigation aims. The research is designed, having two primary aims for the population of interest, in this case new building construction in Germany.

First, to form an objective basis for conceptual estimates, empirical investigation of historical project information is conducted, using two alternative method of analysis, multiple linear regression and artificial neural networks. This incorporates with development of predictive models that can further be employed as an objective ground for practitioners.

Second, to increase prediction accuracy of conceptual estimates to an extent, a novel solution, *multi-way* approach, is adopted along with traditional *one-step ahead* approach. It is hypothesized that estimating building element quantities and later employing them as additional inputs along with meagre information through establishing the basis of a project, multi-way approach may influence a *substantial* decrease in prediction error when compared to conventional, one-step ahead approach. Four alternative frameworks of cost estimation and two alternative frameworks of duration estimation are considered, adopting alternative methods of analysis.

Results reveal that adopting multi-way approach, approximately 19% improvement in prediction accuracy can be accomplished over one-step ahead approach, when the aim is predicting cost of structure. The increase in prediction accuracy is 10%, when duration estimation is considered. However, it is underlined that formal statistical test denies the differences as *substantial* at 5% significance level due to insufficient size of the test sample. Results of analysis also demonstrates, linear regression models provide slightly smaller prediction errors than neural network models. What is more to the point, linear regression models offer significantly lower disperse in error of predictions; and therefore preferred over neural network models. Lastly, confidence intervals of linear regression estimations are computed on a case project to demonstrate application of range estimates. For predictions, like the ones relevant to this research, the possible prediction error inherent in the process itself and therefore the security of the prediction has to be kept in mind.

The models developed within the content of this research relies on objective assessment and scientific methodology and therefore can be regarded as highly reliable for practical implementation by German architects aiming to predict cost of structure and construction duration through establishing the basis of a project. New insights to the German architects are offered by proposing alternative relevant factors in developed models. These factors can be taken into account to predict, monitor and maintain budget and schedule. More to the point, along with observed increase in prediction accuracy compared to conventional practice, employing multi-way approach German architects are able to provide expected average values of building element quantities and hierarchical cost groups according to relevant German standards.

Keywords: Germany; Modeling; Predictions; Cost of structure; Construction duration; Linear regression; Artificial neural networks; Multi-way forecasting

Zusammenfassung

Baukosten und Ausführungsdauer sind neben der Qualität zwei entscheidende Einflussfaktoren für Bauprojekte. Die objektive Schätzung dieser Faktoren ist daher bei der Grundlagenermittlung eines Projekts unabdingbar, um als Basis für Budgetierung, Planung, Ausführung, Bauüberwachung und sogar Rechtsstreitigkeiten zu dienen. Die vorliegende Arbeit untersucht Neubauprojekte in Deutschland als entsprechende Datengrundlage mit zwei vorrangigen Zielen.

Um eine objektive Grundlage für frühzeitige Schätzungen aufzubauen, werden Informationen von fertiggestellten Projekten empirisch untersucht. Dabei kommen zwei alternative Analysemethoden zum Einsatz, die multiple lineare Regression und künstliche neuronale Netze. Daneben werden Prognosemodelle entwickelt, die als objektive Grundlage in der Praxis Anwendung finden können.

Um die Vorhersagegenauigkeit der frühzeitigen Schätzungen möglichst zu erhöhen, wird als neuartige Herangehensweise ein mehrstufiger Ansatz neben dem traditionellen einstufigen Ansatz verfolgt. Es wird angenommen, dass der mehrstufige Ansatz, dass heißt die Schätzung der Elementmengen der Gebäude und anschließende Verwendung dieser Ergebnisse als zusätzliche Inputfaktoren neben einigen weiteren Informationen bei der Grundlagenermittlung eines Projekts, im Vergleich zum konventionellen einstufigen Ansatz eine erheblichen Verringerung des Prognosefehlers bewirken kann. Vier alternative Schemata für die Schätzung der Baukosten und zwei alternative Schemata für die Schätzung der Ausführungsdauer werden dazu anhand zweier alternativer Analysemethoden betrachtet.

Die Ergebnisse belegen, dass bei Anwendung des mehrstufigen Ansatzes im Vergleich zum einstufigen Ansatz bei der Prognose der Bauwerkskosten eine um etwa 19% höhere Vorhersagegenauigkeit erreicht werden kann. In Bezug auf die Abschätzung der Ausführungsdauer beträgt die Erhöhung der Vorhersagegenauigkeit etwa 10%. Es wird jedoch betont, dass ein formaler statistischer Test diese Differenzen aufgrund der unzureichenden Größe der Teststichprobe bezogen auf das 5% Signifikanzlevel nicht als signifikant bewertet. Die Ergebnisse der Analyse zeigen auch, dass die Modelle der linearen Regressionsanalysen geringfügig kleinere Prognosefehler liefern als die der neuronalen Netze. Wesentlich ist weiterhin, dass die Modelle der linearen Regressionsanalyse eine erheblich geringere Streuung beim Vorhersagefehler liefern und aufgrund dessen den neuronalen Netzen vorgezogen werden können. Abschließend werden die Konfidenzintervalle der linearen Regressionsanalysen anhand eines Beispielprojektes errechnet, um die Anwendbarkeit der Bandbreitenschätzung zu veranschaulichen. Bei den Vorhersagen dieser Forschungsarbeit darf der mögliche prozessimmanente Vorhersagefehler und daraus folgend die Zuverlässigkeit der Vorhersage nicht außer Acht gelassen werden.

Die innerhalb dieser Forschungsarbeit entwickelten Modelle stützen sich auf objektive Datenerhebungen und wissenschaftliche Methoden und können daher als sehr zuverlässig für die praktische Umsetzung durch deutsche Architekten betrachtet werden, wenn es um die Abschätzung der Bauwerkskosten und Ausführungsdauer im Rahmen der Grundlagenermittlung eines Projekts geht. Durch die Darstellung alternativer, relevanter Einflussfaktoren in den entwickelten Modellen werden den deutschen Architekten neue Erkenntnisse geboten. Die Faktoren können für die Planung, Steuerung und Kontrolle von Kostenbudget und Zeitplan herangezogen werden. Wichtiger noch, neben dem ermittelten Anstieg der Vorhersagegenauigkeit im Verhältnis zur herkömmlichen Praxis, sind deutsche Architekten bei Anwendung des mehrstufigen Ansatzes in der Lage, die zu erwartenden Durchschnittswerte für Elementmengen von Gebäuden und hierarchischen Kostengruppen nach den einschlägigen deutschen Normen zu ermitteln.

Schlüsselwörter/Stichwörter: Deutschland; Modellierung; Prognose; Bauwerkskosten; Ausführungsdauer/Bauzeit; lineare Regression; künstliche neuronale Netze; mehrstufige Vorhersage

1 Introduction

1.1 Research Background

> *"The physical substance of a house is a pile of materials assembled from widely scattered sources. They undergo different kinds and degrees of processing in large numbers of places, requires many types of handling over periods that vary greatly in length, and use the services of a multitude of people organized into many different sorts of business entity."*

Cox & Goodman (1956) underlined characteristics of the construction industry approximately 60 years ago in a widely known study of the distribution of house building materials. One of the main conclusions is that the number of possible permutations and combinations of specific places and entities is numerous, even for one product. Similarly, the complexity of the industry have been emphasized more recently. Winch (1987) claimed that construction projects are amongst the most complex of all undertakings. Gidado (1996) further remarked and suggested that there is a continuous increase in the complexity of construction projects. The industry's way of functioning and its performance are shaped by these underlying conditions (Dubois & Gadde, 2002). Now and then construction industry participants are blamed for inefficiency of operations (Cox & Thompson, 1997). A number of authors have argued that construction has failed to adopt techniques that have improved in other industries, such as just-in-time (Low & Mok, 1999), total quality management (Shammas-Toma *et al.*, 1998), partnering with suppliers (Cox, 1996), supply chain management (Vrijhoef & Koskela, 2000), and industrialization of manufacturing processes (Gann, 1996). Main properties of construction industry shall be described to reveal why this figure is so.

Cox & Thompson (1997) have argued that construction is inherently *a site specific project-based activity*. Parallel, Shirazi *et al.* (1996) concluded that construction is mainly about coordination of specialized and differentiated tasks at the site level. Gidado (1996), on the other hand, suggested complexity in construction systems caused from interdependence among tasks, and represents those sources of complexity that originate from different parts together to form a work flow. Accordingly, Dubois & Gadde (2002) suggested two central features of construction: (1) the focus on individual projects, in terms of decentralized decision making and financial control, (2) the need for the local adjustment at the construction site. They argued local adjustments are necessary because of the three remaining uncertainty factors: (1) lack of complete specification for the activities at the construction site, (2) lack of uniformity of materials, work, and teams with regard to place and time *(every project is unique)*, and (3) *an unpredictable environment.* Obviously, these characteristic make it difficult or even impossible to apply a centralized approach to decision-making (Dubois & Gadde, 2002).

If construction industry's way of delivering its final product is identified as project-based, than one shall understand the underlying principles of projects. According to Duncan (1996) projects have three main characters: (1) temporary: every project has a definite beginning and a definite end, (2) unique products: a project creates unique deliverables, (3) progressive elaboration: projects develop in steps, and continue by increments. Usually, three main criteria to be controlled in projects are identified as quality, cost (budget), and time (schedule) (Figure 1.1).

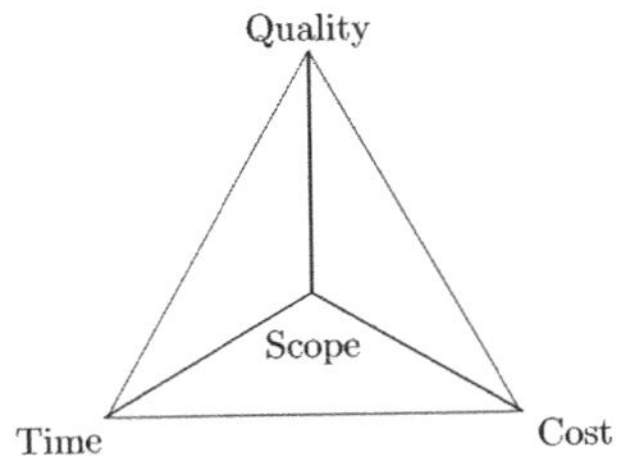

Figure 1.1: Three main success criteria of projects

In this study the major focus is set to investigate a practically applicable solutions for determining budget and schedule of new building projects under high level of uncertainty through establishing the basis of a project. The geographical boundaries of the research is limited to Federal Republic of Germany.

1.2 Problem Statement

Particularly in German practice, it is an *architect*'s responsibility to determine budget and schedule through establishing the basis of a building project. Unlike others, a German architect is authorized for not only design related tasks but also planning and management responsibilities in many cases. There are, of course, exceptions especially when proposed project presents enormous amount of technical and commercial complexities. In these cases, a client may choose to award planning related responsibilities to other professionals in various contractual schemes. Yet, relying on our observations, it can subjectively be argued that the majority of the new building construction follows a traditional route in which the architect is appointed as the head of planning process.

Another aspect of German practice is that it highly relies on standards and regulations regarding building cost classification and identification. Cost planning in building construction, in particular with the identification (forecast) and classification of costs, is defined by DIN 276-1 (2008) norm. Accordingly, building costs are categorized hierarchically in three levels, recognized in three digit numbers (Figure 1.2). In the first hierarchical level of cost classification, the overall costs are divided into seven cost groups. Cost group 300 and 400 can further be combined into a single group called as 'costs of structure' which generally covers a large portion of overall building project costs. Later, first level cost groups are divided into the

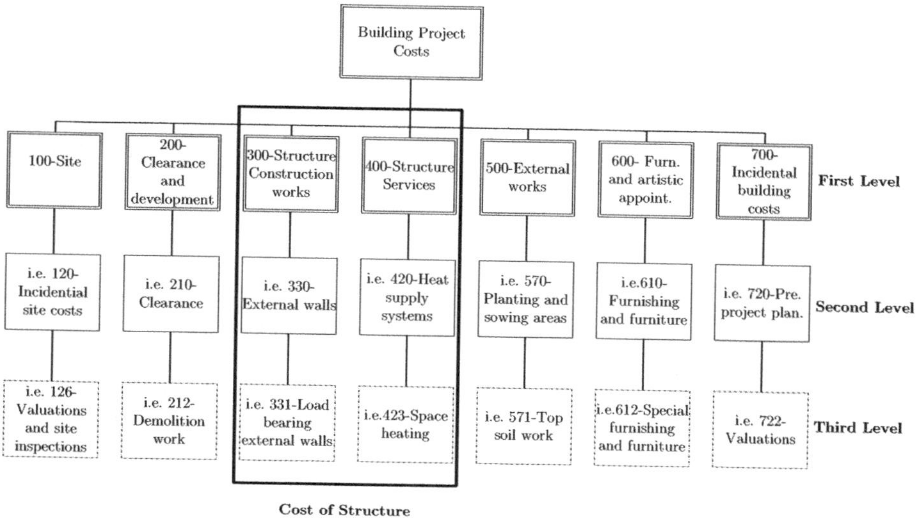

Figure 1.2: Hierarchical structure of DIN 276-1

second and third level cost groups. Another standard, DIN 277-3 (2005), defines reference unit quantities to be employed while identifying the building costs. Cost identification is defined as *"forecasts of costs that will be incurred or determination of costs actually incurred"* (DIN, 2008). Depending on the level of information available to the estimator, cost identification is classified to 5 hierarchical stages:

Table 1.1: Cost groups of structure and corresponding reference quantities (Source: DIN 276-1 and DIN 277-3)

Cost Group	Description (cost of ...)	Reference Quantity
300	*Structure - construction works*	*gross floor area*
310	Excavation	excavation volume
320	Foundation	foundation area
330	External walls	external wall area
340	Internal walls	internal wall area
350	Floors and ceilings	ceiling area
360	Roof	roof area
370	Structural fitments	
390	Other construction related activities	
400	*Structure - services*	
410	Sewerage, water and gas systems	
420	Heat supply systems	
430	Air treatment systems	
440	Power installations	gross floor area
450	Telecommunications systems	
460	Transport systems	
470	Function related equipment and fitments	
480	Building automation	
490	Other services	

1. Budget (*Kostenrahmen*): identification of costs on the basis of demand planning
2. Preliminary estimate (*Kostenschaetzung*): identification of costs on the basis of preliminary planning
3. Approximate estimate (*Kostenberechnung*): identification of costs on the basis of design planning
4. Final estimate (*Kostenanschlag*): identification of costs on the basis of preparation for execution
5. Statement of final costs (*Kostenfeststellung*): identification of the final costs.

Regulation on fees for architects, engineers (HOAI) classifies an architect's tasks to be conducted for a building project into 9 phases (see Figure 1.3). HOAI forwards architects to DIN 276-1 for cost planning through particular project phases (Figure 1.3). In addition, HOAI (2009) suggests a guideline to determine an architect's fee on the basis of cost of structure. This practice is widely employed by architects in Germany. Through determination of budget (*Kostenrahmen*)

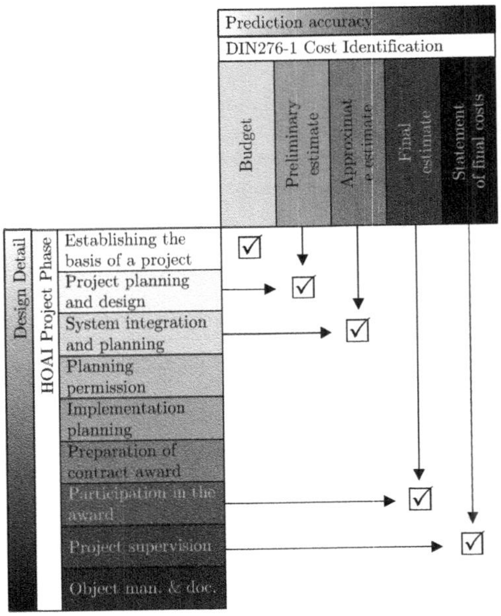

Figure 1.3: Current cost planning practice in Germany (Source: HOAI (2009) and DIN 276-1 (2008))

that is performed through establishing the basis of a project, an architect is, at least, expected to specify costs of structure (sum of cost group 300 and 400) separately within the overall costs (sum of cost groups 100 to 700) (DIN, 2008). In preliminary estimates (*Kostenschaetzung*) which is conducted at project planning and design phase, the overall costs shall be determined by cost groups at least to the first level of the cost classification separately. In approximate estimates (*Kostenberechnung*) that is executed at system integration and planning phase, the overall costs shall be determined by cost groups at least to the second level of the cost classification (DIN, 2008).

Determination of budget and schedule through establishing the basis of a project, which constitutes the primary aim of this research, is crucial; hence Li *et al.* (2005) stated decision making, initial appropriation and economic feasibility studies are based on preliminary estimates that is performed during the early stage of a construction project. Kouskoulas & Koehn (1974) underlined that preliminary estimates lacking a rational basis may prove detrimental to the decision maker and the owners financing a project, in the absence of design plans and specifications. Following, Skitmore & Ng (2003) remarked accurate forecast of construction time and cost is crucial to contract administration as the predicted duration and cost form a basis for budgeting, planning, monitoring and even litigation purposes. Bordoli & Baldwin (1998), on the other hand, stated that the construction industry consistently suffers from project delays and has a relatively poor record regarding completion of projects on schedule. Recent official statistics by the United Kingdom construction industry key performance indicators - industry performance report (2011) clearly stated problem: Only 59% of the housing projects were completed on or under their budgeted price. The figure is very much same for predictability of construction duration: 60% of the housing jobs were executed within their scheduled duration. To our knowledge, there is no official statistics or empirical study undertaken to assess the difference between planned and actual values of cost of structure and duration in Germany. Therefore, average amount of departure between the actual values incurred in the construction site and planned through establishing the basis of a project remains unknown.

In German practice, budget determination, that is performed by an architect through establishing the basis of a project, relies on single-factor calculations (recall Figure 1.3). The single-factor is generally defined as cost (€) per gross external floor area (GEFA) determined by the facility type such as multi family house, offices, etc (Stoy *et al.*, 2008). The problem occurs when the error rate in such forecast is unacceptably high. Ruf (2003), relying on his expertise in cost planning of German building projects, claimed that average error of prediction in cost of structure through establishing the basis of a project is as high as ±30% of the actual final costs. One clear reason of this adverse situation may be the underlying problems incorporate with building construction that was outlined in Section 1.1. Another reason may be lacking of architect's expertise in relevant domain. Third reason may be a fluctuation in the construction market or in the global economy. Time pressure applied by the clients to the architects to conduct and issue early estimates of budget and schedule may also reduce the quality of the task. Client's late selections of design parameters offered by the architect may also be a major reason. What ever underlying reason may be, it can be concluded that early estimations of budget and schedule that is performed through establishing the basis of a project are challenging tasks, generally subject to high error in accuracy. To us, the major problem is able to provide predictions for cost of structure and duration for some confidence through establishing the basis of a project; hence current improvements in information technologies, such as building information modeling is able to provide real time computation of construction costs and duration when design information emerges.

Numerous attempts have been made to enhance precision of determining project targets at the early phases (see Chapter 2). Consequently, many alternative techniques to current estimation practices have been developed. Yet, adoption of new methods has to deal with tremendous amount of resistance by the industry practitioners. Research by Fortune & Lees (1994) showed primary reason is practitioners find it complex to understand how to use them. According to Ross (1998) another problem is the new estimation techniques do not tend to favor incorporating with subjective experience which practitioners find it as a threat to their professional career in longer term. Thus the industry is forced to rely on traditional estimation methods by its own practitioners which has led to dissatisfaction among construction clients with the early estimation techniques (Fortune & Lees, 1994).

In conclusion, German standards, working in collaboration, provide a sound hierarchical framework for cost estimation practices of buildings in Germany. Yet, two major problem remain:

- First, construction cost and duration estimations are performed subjectively. No objective ground has been adopted by the practitioners.
- Second, high error rate due uncertainty and subjectivity is present in determination of the predictions performed at conceptual design phase.

These issues form the main research questions that are addressed in this research for the population of interest, new building construction in Germany:

1. Is it possible to recognize patterns which in turn may form an objective basis for conceptual estimates? By other words, can scientific methodology aid us to develop functional relations between scarce information available through establishing the basis of a project and our targets, cost of structure and duration?
2. If pattern recognition can be reached to an extent, than can developed functions offer superior prediction accuracy when compared to current practices?

1.3 Scope, Aims, and Hypotheses

1.3.1 Research scope

The first restriction on the scope is applied to the geographical boundaries of the study, as mentioned before. The research is limited to Federal Republic of Germany. Second, In Germany, construction works in building projects classify into different types, such as renovation, refurbishment and new construction. Scope of our investigation is restricted to new construction works. Third restriction, on the other hand, associated with *time-point* of estimations. Newton (1991) offered *time-point* of estimations, which implies a limitation on data to an extent, is one of the main categorization criteria for prediction models. The question of which HOAI project phase are the study results intended to use shall be addressed. The scope of the study is strictly limited to the first HOAI project phase - establishing the basis of a project. This implies only macro project determinants that are available to an architect at this stage of the development can be utilized as inputs to generate desired outputs (Chapter 4). Accordingly, the question arises: "How can prediction accuracy be increased if same level of information is utilized?". Whether, it may or may not is, of course, a matter of empirical investigation. Relying on empirical evidences of pilot study, an assumption is emerged that it may be possible.

1.3.2 Research aims

In consistent with the properties and problems, remarked in section 1.1 and 1.2, this study, above all, has two primary aims:

1. To form an objective basis for estimates of cost of structure and construction duration that are conducted by the architect through establishing the basis of a project.

2. To increase the precision of forecasts for cost of structure and construction duration which are performed through establishing the basis of a project.

Construction economics presents a very rich literature on estimating project determinants through establishing the basis of a project. So, determination of appropriate methodologies to investigate the primary research aims are not hard to identify (Chapter 2). However, practical implementation of the study results by the German architects can only be possible, in case frameworks, along with structure of current standards, are offered.

1.3.3 Frameworks and hypotheses

Pilot analysis of the data set demonstrated that macro project determinants are strongly related to the reference unit quantities (RUQ), such as external walls area, of second level cost groups (CG) of structure (see Chapter 4 Section 4.4).

If macro project determinants are precisely available through establishing the basis of a project, than this implies that prediction of reference unit quantities of second level cost groups with some confidence can also be possible with the help of appropriate method of analysis. Later, estimated reference unit quantities can enter as additional inputs to the information system to accomplish the target of estimating cost groups of structure and construction duration in greater precision. Hence more detailed information is revealed and therefore available to an architect, an increase in prediction accuracy, to an extent, can be expected. This iterative process is called *multi-way approach.*

A fair comparison must be addressed to test main hypothesis of reaching superior prediction accuracy in cost of structure and construction duration estimations by adopting multi-way approach. Accordingly, *one-step ahead approach* is considered for comparative aims. In one-step ahead approach, an architect is limited to employ scarce information, so called macro project determinants, to perform conceptual predictions of the values of interest, in this case cost of structure (so called construction cost) and construction duration.

Four frameworks (Figure 1.4) are offered to forecast cost of structure:

- In Framework #1, the process is initiated with estimation of reference unit quantities by using macro project determinants which are assumed to be available to an architect through establishing the basis of a project, HOAI phase number one. In the first step, the aim is to estimate reference unit quantities by employing the macro determinants of the project that are available to the architect through establishing the basis of the project, HOAI project phase number one (see Chapter 4, Section 4.2 for restrictions particular to German practice and available variables). Later, actual values of these quantities in the data set are replaced by the fitted (predicted) values, which are derived

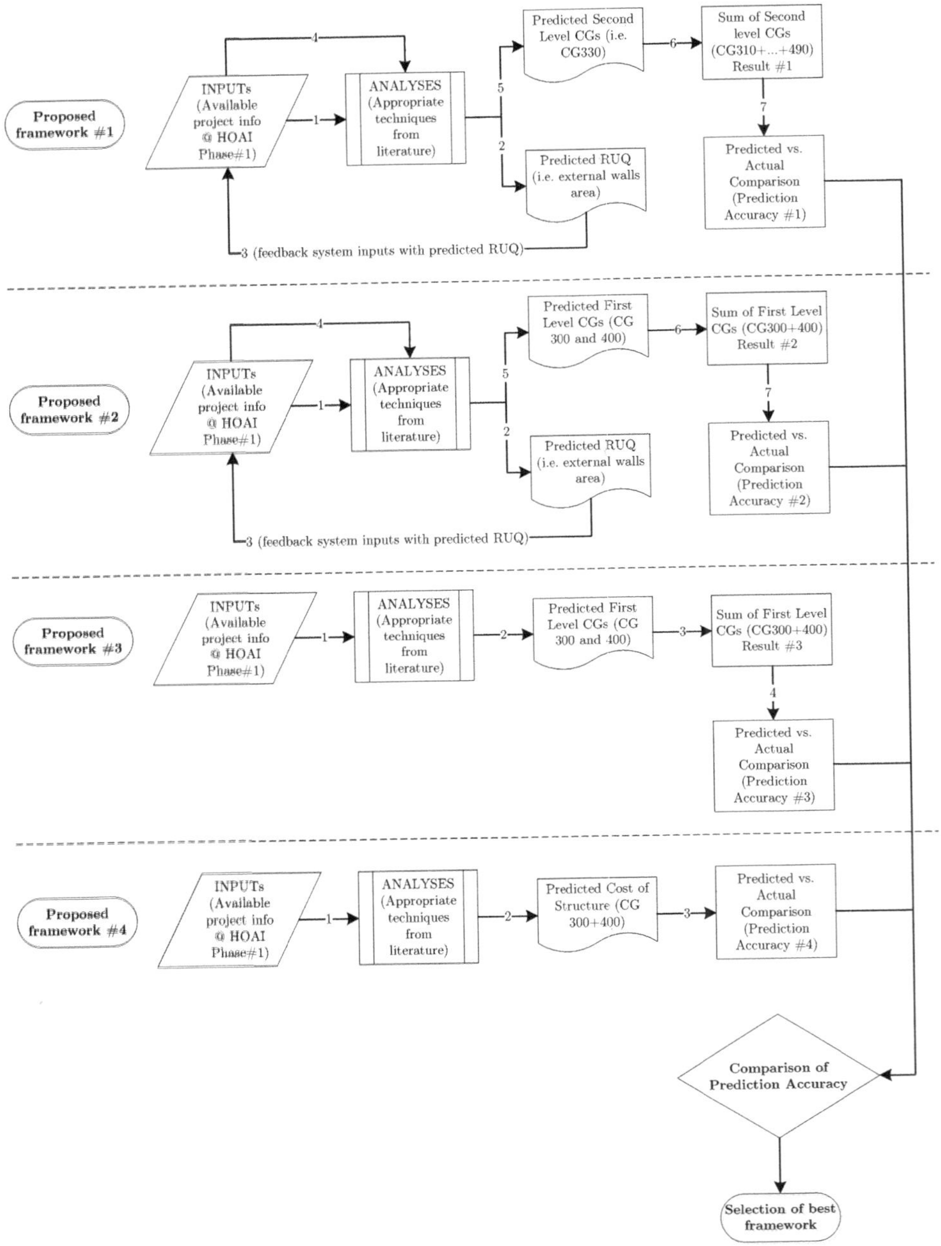

Figure 1.4: Proposed frameworks for cost estimation

by alternative methods of analysis. Two alternative methods of analysis are selected as appropriate tools of forecasting (Chapter 3). The procedure is followed by entering these quantities as additional inputs to the information system. These quantities are considered as additional information along with available macro determinants. Second level cost groups of structure, such as cost of external walls, are predicted with the help of employing best combination of collective information - fitted values of reference unit quantities along with macro determinants. Aggregating predicted values of second level cost groups, predicted cost of structure, as the main target value that will be employed for comparative aims, can be reached.

- It can be argued that using predicted values of reference unit quantities may not be enough to predict second level cost groups of structure at desired level of precision. In this regard, Framework #2 which aims at predicting first level cost groups of structure with multi-way approach is offered. It, again, commences with developing prediction models of reference unit quantities with the limited information available through establishing the basis of a project. In consistent with, multi-way approach, these values are added to information system as additional inputs. The difference between Framework #1 occurs when targeted cost groups are considered. Framework #2 aims at predicting first level cost groups of structure, namely cost of construction works and services (k300 and k400). Aggregating predicted values of first level cost groups, predicted cost of structure, as the main target value that will be employed for comparative aims, can be reached.

- Framework #3 and #4 are developed for comparison aims. Their main role are to test whether any differences in prediction accuracy would occur in between multi-way and one-step ahead approaches. Limited information are utilized as inputs to perform estimations of first level cost groups of structure in Framework #3, and directly aggregated cost of structure in Framework #4.

It is hypothesized that prediction error in cost of structure can be reduced to an extent hence additional information is predicted and later used as inputs.

Accordingly the null hypothesis of:

H_0 = no significant differences observed between prediction error in cost of structure estimation frameworks

will be tested against the alternative hypothesis of

H_{alt} = significant differences observed between prediction error in cost of structure estimation frameworks

Two frameworks (Figure 1.5), Framework #5 and #6, are offered aiming to predict construction duration:

- Framework #5 adopted multi-way approach. Framework #5 can be imagined as an extension to Framework #1 with additional set of predictor variables, namely fitted values of reference unit quantities and second level cost groups of structure. Similar to Framework #1, first reference unit quantities are estimated with the macro project determinants available through establishing the basis of a project. These quantities enter the information system as additional input variables when the aim is to estimate second level cost groups of structure. Once estimated cost groups and reference unit quantities are gathered, these cumulative set of input variables are employed to model and predict construction duration.

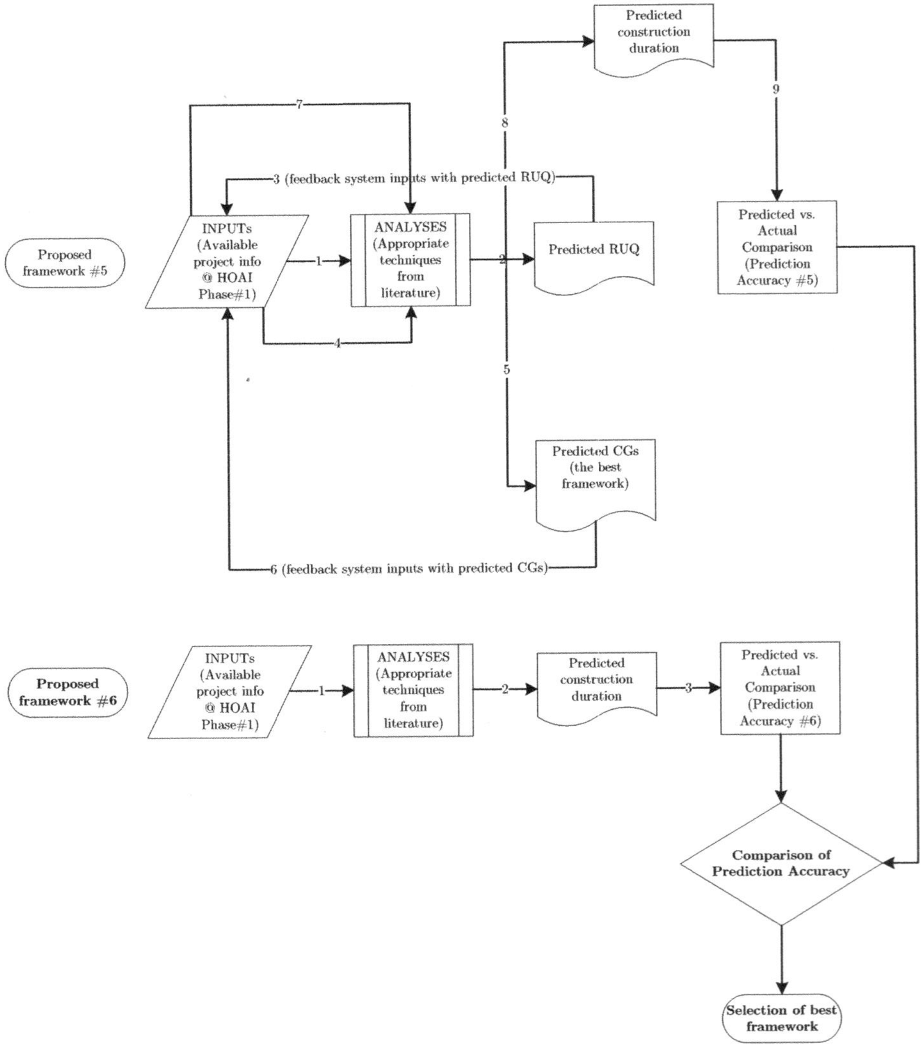

Figure 1.5: Proposed frameworks for construction duration prediction

- Framework #6, on the other hand, adopts one-step ahead approach in which input variables are restricted according to the assumption of this study. Recall, the inputs are limited to a small set of information, macro project determinants, available through establishing the basis of a project.

Similar to cost estimation, it is hypothesized that prediction error in construction duration can be reduced to an extent hence additional information is predicted and used as additional inputs.

Accordingly, the null hypothesis of

H_0 = no significant differences observed between prediction error of construction duration frameworks

will be tested against the alternative hypothesis of

H_{alt} = significant differences observed between prediction error of construction duration frameworks

Received criticism by practitioners Framework #1, #2, and #5 are developed to offer a novel solution to predictions through establishing the basis of a project. This involves in estimation of reference unit quantities of second level cost groups according to DIN 277-3 (2005). This is an ambitious aim, incorporates with predicting expected average quantities of design related information at the first HOAI phase, establishing the basis of a project.

To interviewed architects, predicting design related information is of no practical relevance hence they advocate that design of a building may be subject to large amount of variation through establishing the basis of a project. They conclude their argument by stating no generalization rules can be generated for buildings, at this stage of a project. Ironically, they complain about the uncertainties inherit to building sector and therefore how they fail to employ objective grounds to determine budget and schedule. In addition, they mention of criticisms they have received from their clients, since they fail to adopt scientific methods to their estimations. Yet, an alternative solution was not proposed to improve the current practices.

1.3.4 Research objectives

According to the scope, aims and frameworks, this study can be evaluated in three parts. In the first part of this thesis, models of reference unit quantities, second and first level cost groups of structure, and construction duration are intended to develop. Selected method of analysis (Chapter 3) are the instruments of model development. The first part addresses the primary aim of forming an objective basis for predicting cost of structure and construction duration. Discussions regarding individual models are intended to be conducted in the first part of this manuscript. It is important to emphasize that development of all types of the models are planned to be conducted by using exactly the same number of projects, same data sets and same procedures. This enables to form a fair basis for comparative aims. Selection between two alternative models for a value of interest is intended to be based on prediction accuracy measurements (Section 3.3.2.6).

In the second part of the study, it is planned to conduct a formal test of the hypotheses under investigation. This involves in aggregation of individual model results in accordance with offered frameworks of cost of structure and construction duration estimation for each method of analysis. Motivations stand to select one of the alternative frameworks are intended to be discussed in this part. Therefore, the second part tests whether the primary aim of reducing error of predictions performed at the conceptual design phase can be accomplished, to an extent.

The last part is allocated to exhaustive demonstration of practical implementation of the results using a real building project, that is not previously employed for modeling aims. Predictions are intended to be performed using developed models and selected frameworks. Moreover, computation of range estimations is addressed at the last part.

Accordingly objectives can be listed as:

- In the first part:
  - to determine relevant predictor drivers of reference unit quantities, second level cost groups, first level cost groups, and construction duration. This objective involves in developing predictive models for the values of interests.
  - to compare prediction accuracy between method of analysis to be employed for individual models.
- In the second part:
  - to compare prediction accuracy between alternative frameworks.
  - to perform a formal test of hypothesis.
- In the third part:
  - to demonstrate practical implementation of the resulted models in accordance with the selected frameworks.
  - to compute range estimations

1.4 Structure of the Manuscript

Chapter 1, contains information regarding general properties and issues in building construction. This leads to statement of problems specific to the boundaries of this investigation. Lastly, aims and solutions are offered in consistent with the scope of the study.

Chapter 2 focuses on review of the relevant literature. Those of studies which have addressed the similar issues can be tracked in associated sections. The results of relevant studies are critically discussed. Major differences between our investigation and studies reported are underlined.

Chapter 3 describes the research methodology and methods of analysis that are employed to execute the research, exhaustively. This involves in elaborate demonstration of empirical research methodology, and alternative methods of analysis employed.

Chapter 4, describes the sample employed throughout the research. Descriptive statistics are presented. Assumed available variables (inputs) through establishing the basis of a project, are also listed in detail in this chapter. Lastly, results of pilot analysis are demonstrated briefly to support the ground for hypothesis of interests.

Chapters 5 is allocated to report the findings of model development procedure. Results by individual models are discussed. Comparison of prediction accuracy with respect to alternative method of analysis is performed.

Chapter 6 test the hypothesis of interest. This involves in comparison of offered frameworks of cost and duration estimation. Motivations stand to select one of the alternative frameworks are discussed in this chapter.

Chapter 7 demonstrates practical implementation of the results using a real building project, that is not previously employed for modeling aims. Computation of range estimations are addressed in this chapter.

Chapter 8 presents the main findings of the study.

2 Literature Review

This chapter is solely designated to demonstration of the relevant literature. The main focus through literature review is set along with the scope of the study, as indicated in Section 1.3. This implies that studies undertaken to provide predictions after conceptual stage of a project is not subject to our investigation. Two exception remains for the limitation: (1)readers may also find works that show methodological relevance to our study, (2) review of the literature is not bounded to geographical borders of this study.

Although a great deal of effort has been spent to investigation and determination of relevant works, we fail to detect a research that primarily aims to estimate building quantities when limited project information emerges. Only US Patent number 6,324,508 B1, authored by Finney (2001), makes an attempt to estimate quantities necessary to perform cost estimation subsequent to availability of design documents. Not only, time-point of estimation, but also definitions of quantities presents disparities to the content of our study. Last but not least, Finney's work (2001) is a patent and does not include any hints regarding the methodology employed. On one hand, this causes a challenge to support the ideas and assumptions of our work. On the other hand, this clearly implies the originality and significance of the arguments that is presented in this study.

Despite lacking sufficient number of works that can be reached for reference unit quantity estimation, the figure dramatically changes and literature presents numerous publications that aims at predicting construction cost and duration. For instance, Newton's study (1991), conducted meta literature review that aims to provide an agenda for cost modeling research, cited 74 studies that are relevant to construction cost estimation practices conducted before 1991. Since then, the number of works that have been carried out, has been growing exponentially. This is due to increasing number of related journals, growth in recognition of construction management and economics as a scientific discipline, and advances in computer technology that allows researchers try and implement different methods of analysis.

To sum up, this chapter demonstrates the findings in two section. The first section intends to summarize the relevant studies in construction cost estimation. The second section, on the other hand, targets to outline the works carried out not only to provide construction duration predictions, but also define the relevant drivers of construction duration. In both sections, the main focus is set to provide information regarding primary aims, properties of employed sample, study location, analysis techniques, and main findings. Limitations of relevant study results are also discussed.

2.1 Construction Cost Estimation

2.1.1 Overview

Construction cost estimation has a traditional history in construction economics research. In his well known study, Newton (1991) conducted a meta literature review and, above all, he proposes a taxonomy for the previously developed models. Later, he grouped relevant studies under these categories and suggests to define further work under proposed taxonomy for standardization. He referred the main categories as descriptive primitives. Newton (1991) proposed to categorize descriptive primitives under following nine headings: (1)*Relevance* - whether it relates to a specific design proposal, (2)*Units* - the units of measurement, (3)*Cost/price* - how the model is intended to be used, (4)*Approach* - the level at which modeling is applied, (5)*Time-point* - when during the design process it is applied, (6)*Model* - a general classification of technique, (7)*Technique* - the specific classification, (8)*Assumptions* - whether they can be accessed or not, and (9)*Uncertainty* - how it is treated.

Aims of our research clearly implies time-point of the estimation models to be developed. In addition, state of art in cost estimation research shall be up to date. In this regard, selection of the key studies are based on: (1)According to the year that the study was conducted (focus on works in the last decade), and (2)time-point relevance. It is anticipated that literature offers a range of vital information, especially regarding type of analysis, to shape final context of this research.

2.1.2 Key Studies

Sonmez (2004, 2008, 2011) Sonmez conducted series of studies to assist pre design, so called conceptual, cost estimation of building projects (Sonmez, 2004, 2008, 2011). He employed the same sample consists of care retirement community residential projects that are constructed in States. His early work (Sonmez, 2004) primarily aimed to compare prediction accuracy of linear regression models (LRM) and artificial neural networks (ANNs). He analyzed 30 building projects. Comparison between two techniques was carried out on the basis of mean squared error (MSE), and mean absolute percentage error (MAPE). The sample was separated in two groups: one to perform parameter estimation or train, so called training sample; one to test accuracy of developed models, test group. His approach was first to develop parsimonious regression model, therefore determine the best set of predictor variables; later to investigate whether predictive ability can be improved via employing the same predictors using ANNs. His results indicate that time index, total building are, location index, percent health center and commons area, and the area per unit as the significant regression variables. The regression model accounts for 95% variability in the construction cost. Later, two ANNs with different number of neurons in hidden layer were developed. The results show no significant difference in predictive ability between LRM and ANNs, still favor ANNs slightly over LRM. One important contribution was to emphasize necessity to calculate prediction intervals. (Sonmez, 2004) advocates, due to unfinished scope and limited design information, range estimations show of practical relevance when compared to point estimations.

Following, Sonmez (2008) focused only LRM as the analysis technique to provide parameter based estimation of conceptual construction cost. He downsized sample sample to 20 objects, constructed between 1983 and 1995. He suggested a bottom to top approach and developed models for 11 building cost components including, but not limited to, site development, structure, and interior finishes. 14 available predictors show significant linear relationships with, at least, one cost component. The significant contribution of this study was to present bootstrap sampling method in order to calculate range estimations. This implies combining parametric and probabilistic approaches. Lastly, Sonmez (2011) applied bootstrap sampling into the neural network predictions. Same bottom to top approach is adopted. Bayesian regularization method is used to constrain the size of the network parameters. By doing so, the network can take noise in the data into account.

In conclusion, Sonmez's works are highly relevant to our investigation. Yet, issues related to sample remains. The sample consists of only one type of facility that have been executed by one contractor. This, highly, limits generalization of conclusions. Besides, there are differences between main assumptions. According to Sonmez (2008; 2011; 2004) there are more information available to the estimators at the conceptual estimation when compared to Germany. Still, techniques of analysis and general methodological approach of his works enhances this study's vision.

Stoy (2007, 2008, 2012) Similar to Sonmez, Stoy conducted a series of studies aim at predicting construction cost of building projects at the early project phases (Stoy, 2007; Stoy *et al.*, 2008, 2012). His sample comprises 75 residential building projects that were executed in Germany. Multiple linear regression was selected to explain variability in the response variables. Hence the study was undertaken solely for German residential projects, Stoy (2007) performed cost estimations at two levels in accordance with the hierarchical structure of DIN 276-1. Following, Stoy *et al.* (2008) conducted another study solely focusing on cost of structure, construction works (recall figure 1.2). Six macro project determinants are found to be significant drivers: (1)ratio of external wall area to GEFA, (2)number of elevators, (3)GEFA, (4)construction duration in months, (5)ratio of openings to external wall area, and (6)regional factor. Lastly, Stoy *et al.* (2012) offered an elemental cost model for external walls as an example. The results pointed four variables account for substantial variance in the cost of external walls element: (1)compactness, (2)percentage of openings, (3)the height of the building, and (4)cladding material.

Stoy's work (2007) is influential, hence it provided a bottom to top cost estimation framework in accordance with German norms. Yet, there is a space for further improvement due to selected methodology and limited sample.

Kim et al. (2004) Kim *et al.* (2004) aimed to compare the prediction accuracy of construction cost estimating modeling techniques. The study remarks three modeling techniques to be compared subsequent to a literature review: Regression analysis, artificial neural networks and cased-based reasoning (CBR). The collected data used in that study were the actual construction costs of 530 projects of residential buildings that were built by general contractors between 1997 and 2000 in Seoul, Korea. Performance evaluation of different techniques is considered in terms of Mean Absolute Error Rate (MAER). The models consist of input variables such as gross floor area, number of storeys, number of total units, construction duration in months, roof types, foundation types, basement usage, and finishing grades.

The regression model, best neural network model, and case-based reasoning model with 40 test observations gave MAERs of 6.95, 2.97, and 4.81, respectively. Analysis of variance statistical procedure is utilized to reject the null hypothesis of there are no substantial difference between prediction accuracy of three modeling techniques. The results of the study suggests best neural network model offers superior prediction accuracy when compared to others. Yet, the authors underlined issues related to obtaining the best neural network model architecture and parameters. Furthermore, they remarked that practical usage of neural networks is challenging due to large training and testing times are required when the data is updated. As a conclusion the authors advocate CBR over neural networks and regression analysis for the reason that CBR can also be utilized for explanatory aims which is not possible in ANNs.

Emsley et al. (2002) Emsley *et al.* (2002) aimed to adopt neural network technique to construction cost estimation. Not only neural network but also linear regression models are developed to establish an objective benchmark of prediction accuracy. Necessary empirical data to develop cost models was collected from 288 British building projects. Candidate relevant determinants of total construction cost (Table 2.3) was chosen subsequent to a pilot study.

One of the most significant contributions of this study is the raised discussion associated with the representation of cost as an output. Many studies in the literature employed raw cost as the response variable of developed models. Emsley *et al.* (2002) remarked the assumptions implicit in this choice of variable. Instead, they offer logarithm transformed cost, cost per square meters, and logarithm transformed cost per square meters variables as the outcomes. The authors advocate using cost per square meters provides a measure of cost that is essentially independent of building size. This is to say, understood relationship between raw cost and size of building has been removed from the model.

Modeling procedure initiates with linear regression analysis. In order to obtain the best subset of variables that explains the substantial variance in the response variable backwards and forwards elimination methods were utilized. The results point five significant predictor variables that is included in every linear model. These are gross internal floor area, function of the building, construction duration, mechanical installations, and type of piling if any. Linear regression model results of Emsley *et al.* (2002) are given in Table 2.1.

Table 2.1: Linear regression model results of cost models by Emsley et al. (2002)

	Cost per m2		Log of cost per m2		Log of cost	
	R^2	MAPE	R^2	MAPE	R^2	MAPE
Forward	0.668	20.8%	0.648	20.0%	0.644	20.1%
Backward	0.666	21.7%	0.661	19.3%	0.661	19.3%

Following, neural network models are trained. The initial neural network is fed with five significant cost drivers of regression models. Later, nine of predictors found in five of the six regression models and lastly all variables for which data have been collected are utilized to train neural networks. Due to difficulties and ambiguities involve in determination of network architecture, number of trials, transformation functions, learning rate, training algorithms, and number of neurons in the hidden layer, numerous network models are developed to determine the best combination. Neural network model results of Emsley *et al.* (2002) are summarized in Table 2.2.

Table 2.2: Neural network model results by Emsley et al. (2002)

	Cost per m2		Log of cost per m2	
	R^2	MAPE	R^2	MAPE
Five variable model	0.636	21.9%	0.641	19.6%
Nine variable model	0.688	20.1%	0.685	18.6%
All variables model	0.789	16.6%		
Reduced variables model	0.665	22.7%		

In accordance with the results, Emsley *et al.* (2002) concluded enhanced prediction accuracy of cost estimation when compared to traditional methods such as regression modeling. Yet, they reported that the improvement is not substantial. Still, the research results show strong evidences that non linear relationships between cost per square meters and predictors remain. In addition, it is reported, interactions that can not be taken into account in linear regression modeling may be present and shall be addressed in further work.

Lowe et al. (2006) Using 286 building project information collected in the UK, Lowe *et al.* (2006) aimed at developing conceptual cost estimation models with the help of multiple linear regression analysis. Similar to Emsley *et al.* (2002), the study addressed the issue of representation of cost as an output variable. Lowe *et al.* (2006) argued using raw cost as a dependent variable violates number of assumptions for linear models. Thereof, they reject using raw cost as the response variable for the models to be developed and proposed three alternatives which are logarithm transformed raw cost, cost per square meters and logarithm transformed cost per square meters.

Table 2.3: Input variables of Lowe *et al.* (2006); Emsley *et al.* (2002)

Project strategic variables	Contract form (O)	Duration (S/weeks)
	Procurement strategy (O)	Tendering strategy (O)
	Purpose (O)	
Site related variables	Site access (O)	Type of site (N)
	Type of location (O)	Topography (O)
	Air conditioning (S/£)	Shape complexity (O)
	Function (S/£)	Stories below ground (S/number)
	GIFA (S/m^2)	Mechanical installations (S/£)
	Height (S/m)	Piling (N)
	Internal doors (S/£)	Protective installations (S/£)
	Internal walls (S/£)	Quality (O)
	Internal wall finishes (S/£)	Roof construction (O)
Design related variables	Number of lifts (S/number)	Roof finishes (S/£)
	Stories above ground (S/number)	Roof profile
	Ceiling finishes (S/£)	Special installations (S/£)
	Electrical installations (S/£)	Stairs (O)
	Envelope (S/m^2)	Substructure (O)
	External doors (S/£)	Structural units (S/number)
	External walls (S/£)	Upper floors (O)
	Floor finishes (S/£)	Wall-to-floor ratio (S/ratio)
	Frame (O)	Windows (S/£)

Note: N=nominal (e.g., piling: 0=no piling, 1=piling); O=ordinal (e.g., shape complexity: 0=low, 1=medium, 2=high); and S=scale (e.g., external doors: a proportional scale based on relative costs)

41 candidate predictors were present in the data set (Table 2.3). For each response variable forward and backward regression models were developed. Significant predictors for each model are indicated in Table 2.4. One can see in Table 2.4 that five variables appear in all six models: GIFA, function, duration, mechanical installations and piling. The evidence suggests that these are the key determinants of construction duration when linear relationship is investigated. The highest prediction accuracy in terms of MAPE was obtained from backward regression model for log cost/m^2.

Gunaydin and Dogan (2004) Murat Gunaydin & Zeynep Dogan (2004) conducted a study aimed at utilizing neural network approach of early cost estimation of structural systems of

Table 2.4: Variable included in the regression models of Lowe *et al.* (2006)

	Log cost		Cost/m^2		Log Cost/m^2	
	Forward	Backward	Forward	Backward	Forward	Backward
GIFA (log)	■	■	■	■	■	■
Function (log)	■				■	
Function		■	■	■		■
Duration (log)		■		■	■	■
Duration	■		■			
Mechanical installations	■	■	■	■	■	■
Piling	■	■	■	■	■	■
Internal wall finishes	■	■	■		■	■
Frame	■	■	■		■	■
Site access	■	■	■		■	■
Protective installations	■	■	■	■	■	■
Internal walls		■		■		■
Substructure				■		
Wall/floor ratio		■				■
Special installations				■		
External walls				■		
Floor finishes				■		
Height (log)		■				■
Units		■				■
Electrical installations		■				■

buildings. Their sample comprises 30 residential building projects between 4 and 8 storey height and were executed in Turkey. The study does not cover any other modeling techniques than neural networks. The comparison was conducted on the basis of other works that has been conducted in different geographies. Murat Gunaydin & Zeynep Dogan (2004) chose eight design parameters to predict the cost of the structural system per square meters (Table 2.5).

Table 2.5: Model determinants to predict cost of structural system per square meters (Murat Gunaydin & Zeynep Dogan, 2004)

Definition	Range
The total area of the building	330-3484 m2
The ratio of the typical floor area to the total area of the building	0.07-0.26
The ratio of ground floor area to the total area of the building	0.07-0.30
The number of floors	4-8
The console direction of the building	One direction (0), two directions (1)
The foundation system of the building	Pier (0), wall (1), slab (2)
The floor type of the building	Reinforced concrete (0), precast concrete (1)
The location of the core of the building	At the site (0), middle (1)

A multi layer feed forward network with four hidden neurons is selected as the architecture. Hyperbolic tangent curve function is used as the activation function. Back propagation algorithm best fits the problem of optimizing the network weights between layers. Different than others, Murat Gunaydin & Zeynep Dogan (2004) employed mean square error (MSE) as the measure of accuracy along with MAPE. Last but not least, over-training problem was underlined and a solution to prevent the issue was offered. Accordingly, number of epochs during training were limited to a maximum value of 10,000. Data from six projects were separated as the test sample. The results indicated 93% average accuracy with a MSE of 0.038.

Karanci (2010) Karanci (2010) studied on conceptual cost estimation models for his Master thesis. The study focused on comparison between three modeling techniques, LRM, ANNs and CBR. Last but not least, Karanci (2010) took uncertainty into account and provided range estimations by utilizing bootstrap sampling method.

His sample consists of 41 mass housing projects from 26 different provinces that were constructed in Turkey between 2003 and 2009. He hypothesized 12 candidate predictors would be available to an estimator (Table 2.6). Similar to our approach, overall cost of a building project was divided into six components. Yet, these cost groups were not defined in accordance with any standard but contractor's own classification. Thereof, there is no consensus on the definitions which limits the practical applicability of the results by others.

Table 2.6: Candidate predictors of cost models (Karanci, 2010)

Predictor	Initial model in which the predictor is included
Building construction cost index	All
Project duration (days)	GENR
Total construction area	All
Total site area	ELEC, MECH, INFR
Total number of apartment blocks	ELEC, MECH, INFR, STR&ARC
Total number of apartments	ELEC, MECH, INFR, STR&ARC
Percent area of social, health and educational facilities in the total construction area	ELEC, MECH, INFR, STR&ARC
Earthquake region	STR&ARC
Site topography	INFR
Type of insulation	STR&ARC
Number of elevator stops	CONS
Classification for degree	MECH, STR&ARC

Note: STR&ARC=structural and architectural works, MECH=mechanical works, ELEC=electrical works, INFR=infrastructural works, CONS=conveying systems, GENR=general requirements

Karanci (2010) employed backward regression to develop parsimonious regression models for each cost component. Later, significant predictors were employed to develop ANNs and CBR models. The results demonstrated that LRM reached to smaller value of MAPE and MSE when compared to ANNs and CBR models. In addition, it was shown that parsimonious models provides higher prediction accuracy than complex models.

Last but not least, Karanci (2010) estimated range of three prediction methods on an example case by using bootstrap sampling. The result of the comparison demonstrated lower variability of cost has been established when LRM is considered. This is to say, estimation range of LRM is smaller than those of ANNs and CBR. Thereof, he concluded that the variability in the construction cost of mass housing projects in Turkey can best be explained by LRM.

2.1.3 Conclusion

Detailed review of the cost estimation models that were developed in the last decade was conducted in the last section. It shall be emphasized that the authors are aware of other studies conducted in the last decade. Yet, our research focus those of studies which present similarities to ours in terms of time-point of estimations. Models developed by others that are intended to be employed at different stages of projects were investigated but were not explained in detail in literature review for clarity. Likewise, studies that present no significant originality are also excluded. To us, demonstrated studies shows a clear picture of current agenda in conceptual cost estimation research.

Approaches of statistical methods, particularly linear regression analysis, are one of the most applied techniques to cost estimation practices since 1970s [e.g.(Kouskoulas & Koehn, 1974; Bowen & Edwards, 1985; Khosrowshahi & Kaka, 1996)]. Yet, there are number of limitations implicit to employment of linear regression such as relationship must be explained by the help of a linear function, assumptions regarding the model and error terms and so on (see Chapter 3). In addition, some (Tam & Fang, 1999) discussed that interactions between predictors, which is certainly present for building research, can not be assessed properly.

Table 2.7: Classification of the key studies with respect to method of analysis

Author (year)	Method of analysis		
	LRM	ANNs	CBR
Emsley *et al.* (2002)	■	■	
Sonmez (2004)	■	■	
Murat Gunaydin & Zeynep Dogan (2004)		■	
Kim *et al.* (2004)	■	■	■
Lowe *et al.* (2006)	■		
Stoy (2007)	■		
Sonmez (2008)	■		
Stoy *et al.* (2008)	■		
Karanci (2010)	■	■	■
Sonmez (2011)		■	
Stoy *et al.* (2012)	■		
Frequency	*9*	*6*	*2*

Consequently, an alternative branch of artificial intelligence, neural networks, was introduced to estimate construction cost during 1990s [e.g. (Li, 1995; Adeli & Wu, 1998; Hegazy & Ayed, 1998)]. ANNs are particularly powerful hence their capability of capturing complex interactions between predictor variables, as well as no necessity for pre-assumed relationship between predictor and response variables. Yet, some (Hegazy *et al.*, 1994) argued ANNs are black-box techniques, and knowledge acquisition process is time-consuming. However, it shall be noted that improvement in processing speed, even for personal computers, in the last decade dramatically reduce the amount of time necessary to train ANNs.

Review of the relevant literature suggests followings:

- LRM and ANNs have still been dominant modeling techniques for conceptual cost estimations in the last decade (Table 2.7).

- Research community shows no consensus on superior modeling technique in terms of prediction accuracy. Hence relevant studies are empirical, employed data significantly shaped the results. Besides, numerous variables with various definitions are included in the models. This makes it impossible to compare the results of studies with each other.

- It is evident with the type of candidate predictor variables that the definition of pre design or conceptual estimates presents disparity between different locations. In Germany, the definition is provided by HOAI while in other studies it was left uncertain. As Newton (1991) remarked, this restricts the number and type of variables available to the study. For instance input variables, such as roof profile, ceiling finishes, shape complexity, and floor finishes, assumed by Emsley *et al.* (2002); Lowe *et al.* (2006) (Table 2.3) can not be accessed in German practice at HOAI basic evaluation stage.

- In general, cost models aimed to provide point estimations. This indicates many did not take uncertainty of construction into account. It is very unlikely, for a building to be completed in a way that it is initially imagined by a client and designed by an architect. Variations are inevitable. Thereof, to us, providing range estimations is vital and presents higher practical relevance.

- There is no doubt that relevant works are based on strong theoretical and methodological foundations. Yet, size of the samples plays a crucial role of those studies that aims to generalize a pattern to predict future values. It is noted that many of the studies fail to reach sufficient number of observations for empirical analysis. This implies limitation of inferences to an extend.

In conclusion, LRM and ANNs are selected as the appropriate techniques to analyze the data to be used in this research. Especially, ANNs has never been implemented in Germany to address building cost estimation problem which may provide a novel solution for superior predictive accuracy. Secondly, we argue that uncertainty shall be addressed. Last but not least, there is no global consensus on definition of different project stages. Our restrictions on data are firmer than the others. This obstructs conducting a fair comparison in between the results.

2.2 Construction Duration Estimation

2.2.1 Overview

Modeling duration estimation has been at the center of attraction of many researchers for decades. Many different models were developed in order to guide the industry (Skitmore & Ng, 2003; Bromilow, 1969; Chan, 1999; Stoy *et al.*, 2007; Love *et al.*, 2005; Ogunsemi & Jagboro, 2006). Yet none of them was able to reach the popularity of Bromilow's time-cost model (BTC), which is far more practically applicable than most of the other models. Therefore, the BTC model is chosen as a starting point for the explanation of the theoretical background with regard to the modeling of construction duration for building projects. Moreover, this chapter aims at presenting contra views and criticisms of the BTC model and presents other models that were developed as alternatives to the BTC model.

To achieve these objectives, an exhaustive literature review was carried out and the essence of the relevant studies will be outlined. They were executed in different geographies including Australia (Skitmore & Ng, 2003; Ng *et al.*, 2001; Bromilow, 1969; Love *et al.*, 2005; Ireland, 1985), Hong Kong (Chan, 1999; Kumaraswamy & Chan, 1995; Chan & Kumaraswamy, 1999,

1995), UK (Kaka & Price, 1991), Malaysia (Chan, 2001), Nigeria (Ogunsemi & Jagboro, 2006), Germany (Dursun & Stoy, 2012; Stoy *et al.*, 2007), USA (Hoffman *et al.*, 2007), and Turkey (Dursun, 2008) during the last four decades.

2.2.2 Bromilow's time-cost model and related studies

The first empirical study with the objective of testing the contract time performance was carried by Bromilow in Australia. Bromilow (1969) aimed at modeling construction duration empirically. The study resulted in the development of his well-known time-cost model, which enables the prediction of construction duration according to the estimated final cost of a construction project. Following, Bromilow (1974) surveyed 370 completed Australian building projects in order to find the determinants of the execution phase's duration of construction projects. He suggested that there is one single variable that affects construction duration: the project size. Therefore, he postulated a model where the construction duration is a function of project size, which he measures as the estimated project cost for clients. The central equation is given by

$$T = K \times C^B \tag{2.1}$$

where, T is the duration of the construction phase of a project measured in days (starts from the possession of construction site and ends at practical completion). C is the estimated final cost of the construction phase of a project in millions of dollars which is adjusted to constant material and labor prices. K is a constant describing the general level of time performance for an Australian $1 million project. Finally, B is a constant describing how the time performance is affected by project size as measured by cost. According to his analysis, the relationship between the duration and the cost of Australian building projects can be expressed as:

$$T = 313 \times C^{0.3}$$

Since then, several studies were carried out in Australia to adjust the BTC model [e.g.(Ng *et al.*, 2001; Ireland, 1985)]. The BTC model is a widely accepted standard for predicting the contract duration of construction projects (Ireland, 1985).

Ireland (1985) intended to validate the BTC model with a limited sampling scope. He carried out the study exclusively with high-rise commercial projects in Australia. According to the analysis of the information from 25 completed projects, he concluded that the best predictor of the average construction duration of high-rise commercial buildings is cost. He formulated his model as follows

$$T = 219 \times C^{0.47}$$

Kaka and Price (1991) intended to estimate the correlation between project duration and project cost in the UK by the means of the BTC model. In this context, Kaka and Price (1991) collected 661 building projects and 140 road projects between the years 1984-1989 in the UK in order to validate the BTC model. According to their results, which are given for different categories in Table 2.8 and Table 2.9, even though the estimated and actual values of the projects vary swiftly, the relationship between duration and cost stays the same. In addition, they concluded that the type of the project and the type of the client influence the relationship between duration and cost. Another relevant outcome of the research is that building projects of the private sector varied remarkably (R=0.49) and the model fitted the data moderately.

Kaka and Price (1991) refer to the private buildings sector as high volatile and unpredictable. On the other hand, they suggest that further classification might improve the accuracy of the model and could therefore help solving the problem.

Table 2.8: Model parameters and effect measures for tendered and actual road contracts in England

Civil (roads)	B	K	R
Tendered	0.469	258.1	0.84
Actual	0.432	245	0.84

Table 2.9: Model parameters and effect sizes of buildings categorized by client, contract and tendering type in England

Building category	B	K	R
Public buildings			
Public fixed price contracts	0.318	398.8	0.76
Public adjusted price contracts	0.205	486.7	0.68
Public open competition tendering	0.293	407.4	0.74
Public selected competition tendering	0.342	424.1	0.82
Public negotiated competition tendering	0.372	367.5	0.77
Private buildings			
Private fixed price contracts	0.212	274.4	0.49
Private adjusted contracts	0.082	491.2	0.61

Yeong (1994) investigated the relationship between building duration and building costs of construction projects both in Malaysia and in Australia. Based on compiling data of 67 Australian governmental projects, 20 Australian private sector projects, and 51 Malaysian governmental projects, he concluded that the BTC model fits the data very well. He formulated his outcomes as follows

Australian private projects

$$T = 161 \times C^{0.367}$$

Australian governmental projects

$$T = 287 \times C^{0.237}$$

All Australian projects

$$T = 269 \times C^{0.215}$$

Malaysian government projects

$$T = 518 \times C^{0.352}$$

Kumaraswamy and Chan (1995) carried out series of research projects in which they aimed at describing the determinants of construction duration and modeling duration of public housing projects in Hong Kong, respectively. They delivered a questionnaire to 400 companies and received 111 replies. According to their results, which are given in Table 2.10, they concluded there are many factors affecting construction duration, with the major factors being construction cost, gross floor area, and number of floors. They executed the BTC model in order to assess the relationship between duration and cost. The BTC model fits the data significantly.

Table 2.10: Estimated and actual values and resulting effect sizes for public and private buildings in Hong Kong

Type of building	Estimated			Actual		
	K	B	R	K	B	R
Public building projects	182.3	0.277	0.81	216.3	0.253	0.79
Private building projects	202.6	0.233	0.69	250.9	0.215	0.65

Chan, A. P. C. conducted two different studies to test the BTC model in two different geographies. In his first study (1999), he validated the BTC model in Hong Kong's building industry, whereas his second study (2001) takes place in Malaysia. Chan (1999) collected information about 110 projects, which were completed between the late 1980's and the early of 1990's, for his study. As shown in Table 2.11, Chan (1999) pointed out a significant correlation of time and cost for Hong Kong's building projects as an outcome of his study.

Table 2.11: Correlation and regression results of the time-cost relationships of building projects in Hong Kong

Type of project	K	B	R
All building projects	152.082	0.29161	0.9218
Public building projects	166.257	0.28098	0.9543
Private building projects	119.569	0.33725	0.8533

Chan (2001) executed his second study to investigate whether BTC model can be extended to public building projects in Malaysia. Chan (2001) collected time and cost data from 51 public projects and he formulated the relationship between time and cost as follows:

$$T = 269 \times C^{0.32} \qquad (R = 0.638)$$

Chan (2001) stated that one of the main reasons that motivated him to execute the research in Malaysia is a lack of standardization and objectivity for duration estimation at the Malaysian building industry.

Ng, et. al. (2001) executed their study in Australia in order to assess the predictive ability of the BTC model. When natural logarithm is applied to 2.1, the model takes a linear form:

$$ln(T) = ln(K) + B \times ln(C)$$

The authors found it complicated to use millions of dollars instead of normal dollar units. Therefore Ng, et. al. (2001) defined c=1,000,000C. The study examined 93 building projects in Australia and the analysis resulted in the following relationship between duration and cost of a project

$$ln(T) = 0.5844 + 0.3105 \times ln(C)$$

where $R = 0.7668$; $F = 129.84$; and $p < 0.0000$ Ng, et. al. (2001) also made a comparison for the K and B values between previous publications and the study carried out. It was concluded that a significant decrease in K value took place during 3 decades, whereas the value of B stays at the same levels. The figure reveals that the productivity in the Australian building industry increased remarkably. In addition, the study reported that the contract time performance of 'public' and 'private' sectors do not vary remarkably.

Hoffman, et. al. (2007) carried out a study to estimate the performance time for construction projects in the USA for which facility projects financed by the Air Force with construction work beyond \$750.000 were considered. In total, information on 856 facility projects that were completed between the years 1988 and 2004, was collected and analyzed. Applying the BTC model to the sample data resulted in the following model equation

$$T = 26.8 \times C^{0.202}$$

The overall model was found to be significant with $F = 295$ and $p < 0.0000$. Also the model fitted the population moderately with $R = 0.58$.

Dursun (2008) executed a study in Turkey in order to validate the BTC model for residential building projects in Istanbul. Information from 75 residential building projects were sampled. These project were commenced after the 4th quarter of 2001 and completed before the 4th quarter of 2008. A significant relationship between construction duration (time) and estimated construction cost was obtained with $F = 27.91$ and $p \leq 0.000$ [5]. Applying the BTC model to the sample data resulted in the following equation

$$T = 428.38 \times C^{0.171}$$

The model fitted the data moderately with $R = 0.526$.

Following, Dursun & Stoy (2011) aimed at testing statistical adequacy of categorization with respect to project location based on BTC model. A sample of 246 industrial buildings are collected from a global German manufacturing firm. The result of the study favored full model to the restricted one. This implies that different regression equations with distinct model parameters are present and therefore, the data set consists of two different populations - namely, projects in Germany and projects in other locations. Resulted equations are:

Projects in Germany

$$T = 2.30 \times C^{0.335} \qquad (R^2_{(adj.)} = 0.55)$$

Projects in other locations

$$T = 2.36 \times C^{0.183} \qquad (R^2_{(adj.)} = 0.30)$$

2.2.3 Critics on BTC model and other models

Bromilow's study about the relationship between time and cost is one of the most cited studies for this topic. Moreover, in the scientific context his model is the one which is most frequently applied for the prediction of project duration due to its practical applicability and frequently encountered high accuracy level. Various studies in Australia, the UK, Hong Kong, Malaysia, the USA and Turkey revealed the occurrence of high correlations between project duration and project cost while validating the BTC model.

However, the predictive ability of the BTC model is under investigation for decades. One inadequacy of the BTC model is that it fails to assess the impact of determinants on construction duration other than construction cost (Walker, 1995). Walker (1995) attempted to calculate the relationship between the time performance of a construction project and gross floor area, but it was inappropriate to assess the project scope with gross floor area because of external work components.

Ireland (1985) aimed at improving the accuracy of the BTC model by adding new determinants such as gross floor area and number of stories and applying multiple regressions. However, unreasonably high standard errors halted the progression.

Chan and Kumaraswamy (1999) suggested that not only major factors like project cost, gross floor area etc., but also minor factors such as productivity are correlated with construction duration. In order to be able to model the duration of public housing projects in Hong Kong in their second publication on this topic (Chan & Kumaraswamy, 1999) in their first study Kumaraswamy and Chan (1995) harmonized previous research results and micro determinants that influences the project duration. The research modeled estimated construction duration with the predictors estimated cost, presence of precast facades, height of the building, nature of site, and type of scheme. The authors suggested for future research to carry out a research with larger sample sizes in different categories, investigate for the new determinants, and adopt them to the existing model.

Skitmore and Ng (2003) carried out a study to develop forecast models for actual construction time and cost in Australia. The proposed forecast model can be used if client sector, contractor selection method, contractual agreements, project type, contract period and contract sum are known. Therefore the developed model is only applicable for the post contract stage of the development where all variables can be identified (Skitmore & Ng, 2003). All developed estimation models during their research assume that the contract sum and contract duration are known while in practice these determinants can only be predicted at the estimation phase with the help of available information (Skitmore & Ng, 2003).

According to the research that was carried out in Australia, Love, Tse and Edwards (2005) argued that cost is a poor predictor for the duration of construction projects, since the accurate calculation of construction costs at the initial steps of a project is impossible due to change orders and rework. Consistent with the analysis of sample data, Love, Tse and Edwards (2005) suggested that gross ground floor area (GFA) and the number of floor levels can be used to predict duration of building projects rather than cost. Their research resulted in a duration prediction formula, which can be expressed as:

$$log(T) = 3.178 + 0.274 \times log(GFA) + 0.142 \times log(floor)$$

Ogunsemi and Jagboro (2006) intended to model the time-cost relationship for the Nigerian building industry. First, the study tried to validate the BTC model for the industry. For the analysis of the study, information from 87 completed (between the years 1991-2000) building projects was collected. They formulated their results as follows:

All projects

$$T = 63 \times C^{0.262}$$

Private Projects

$$T = 55 \times C^{0.312}$$

Public Projects

$$T = 69 \times C^{0.255}$$

Even though some of the assessment measures like the F – ratio and root mean square error tend to favor the model, the coefficient of determination (R^2) which is widely accepted as an indication of how well the model fits the population is very low (Ogunsemi & Jagboro, 2006). Therefore, the authors decided to model the time-cost relationship with the piecewise linear model with a break-point (BPT), whose general form can be expressed as:

$$T = a_0 + a_1 \times C^*(C \leq BPT) + a_2 \times C^*(C \geq BPT)$$

The resulting models are:

For all projects

$$T = 118.563 + 0.401 \times C(C \leq 408) + 603.427 + 0.610 \times C(C \geq 408)$$

For private projects

$$T = 169.895 + 0.491 \times C(C \leq 557) + 709.66 + 0.884 \times C(C \geq 557)$$

For public projects

$$T = 98.01 + 0.367 \times C(C \leq 353) + 567.967 + 0.283 \times C(C \geq 353)$$

Ogunsemi and Jagboro (2006) also concluded that the public and private project durations vary for Nigeria. This is in disagreement with Yeoung (1994), and Ng et al. (2001).

Hoffman et al. (2007) reported a correlation between project duration and project cost when validating the BTC model during research carried out in USA for Air Force facilities construction works. However, in order to develop a more accurate estimation model additionally multiple linear regression was applied. This resulted in a more successful model which can be stated as

$$y = 3.44 + 0.198x_1 - 0.059x_2 - 0.07x_3 - 0.222x_4 - 0.193x_5 - 0.0146x_6$$

where y is equal to project construction duration, is equal to project construction cost after natural logarithm transformed, and the remaining values represent dummy variables for additional predictors.

Stoy et al. (2007) executed a study aiming at modeling construction duration of residential building projects in Germany. To achieve the objective, indicators, that make duration estimation possible during the early stages of development, were identified by literature scanning. Data from 115 residential building projects were collected. Simple regression analysis was applied to analyze the data. Construction speed, which is measured as m2 of gross external floor area per month, was selected as the dependent variable of the analysis (Stoy *et al.*, 2007). The study results reveal that project size, measured as gross external area in square meters, and project standard, measured as building construction cost in € per gross external floor area in square meters, are found to be the major influence factors of construction speed for residential building projects in Germany (Stoy *et al.*, 2007). The analysis yielded the following model

$$ln(y) = 4.753 + 0.0002x_1 - 0.001x_2$$

where y is equal to construction speed, is equal to absolute size and is equal to project standard. Therefore the study by Stoy, et. al. (2007) supports Bromilow's (1969) research outcomes for Germany, where project size is found to be a significant determinant of project duration. However, Stoy *et al.* (2007) argued that cost of the project was inappropriate for describing project size in the residential building sector in Germany.

Two exceptional studies are differentiated in terms of employed methodology from the rest in construction duration research. First, Bhokha & Ogunlana (1999) aims to implement ANNs to estimate construction duration at the pre design stage. 136 building project information situated in Greater Bangkok compose the sample. Bhokha & Ogunlana (1999) reached to 18.2% accuracy level by using ANNs. Their results indicate that trained ANNs tend to overestimates more than it underestimates. To our knowledge, work of Bhokha & Ogunlana (1999) is the first application of ANNs of duration estimation in construction economics field.

Second, Dursun (2009) employed survival analysis to model construction duration of building projects in Istanbul. The sample consists of 161 building projects that were constructed between 2001 and 2008. Dursun (2009) indicated that survival analysis deals with time to event data and the most important reason to select such methodology is that time data is always non-negative and therefore positively skewed. This restriction of range violates the assumptions of linear analysis techniques. Due to limited availability of predictor variables, prediction accuracy of developed model is limited. Yet, presenting and implementing an advance statistical analysis shows an original contribution to the construction duration estimation.

2.2.4 Conclusion

Bromilow's studies are a widely accepted, common standard for duration modeling at the early stages of a building project, where few project indicators are known. There is evidence of the validity of his model at various places all over the world.

Bromilow (1969) assessed different determinants that influence construction duration and concluded that project size, estimated by construction cost, is the best predictor. On the other hand, it resulted that project cost presents not only a dimension for the project size but is also a sign of project complexity and quality (Hoffman *et al.*, 2007). In addition to Bromilow (1969), many researchers reported a high correlation between project size and project duration [e.g. (Chan, 1999; Kumaraswamy & Chan, 1995)]. However, different determinants, such as external gross floor area and number of storeys, were taken as approximate measures of project size in some of the studies [e.g. (Love *et al.*, 2005; Stoy *et al.*, 2007)]. Furthermore, some studies tried to assess different dummy variables like procurement method or building type in addition to the major determinants (Hoffman *et al.*, 2007; Love *et al.*, 2005).

It can be criticized that the presented studies on modeling construction duration are lacking advanced methodology as merely simple analysis were applied. Apart from Dursun (2009) and Bhokha & Ogunlana (1999) there was no study that used a method other than simple or multiple linear regression. A consequence might be the insufficient approximation of the real relations between the variables of interest due to restrictions in the chosen statistical approaches. Therefore, it is generally desirable to seek and implement innovative methodology.

3 Methodology

This chapter incorporates with exhaustive demonstration of research methodology and analysis techniques employed to conduct the study. In this context, an attempt to present the basic definitions regarding empirical research was made. Following, detailed explanation of LRM and ANNs was presented. The information given in this chapter address those who are not competent at epistemology of empirical study and also lacking of mathematical background of statistical modeling and neural networks. It is aimed to provide sufficient, brief and clear explanations and definitions, systematically.

3.1 The Empirical Work

3.1.1 A General outlook

The experimental style of research is, perhaps, suited best to bounded problems or issues in which the variables involved are known, or at least hypothesized with some confidence (Fellows & Liu, 2008). An experiment is an activity or process, a combination of activities, which produces events, possible outcomes. Ideally, variables should be isolated through the design of an experiment such that only one of the, possibly, very many independent variables values is changed and the consequences on the isolated single dependent variable is monitored and measured accurately. However, in social investigations, including construction management and construction project based experiments, it is neither practical nor possible to allow only one independent variable to alter in value, nor is it possible to isolate individual dependent variables on most occasions, hence the usual approach to experimental design is to devise a study in which the main independent variables, except the one of interest, are held constant and the consequences for the major dependent variable are measured (Fellows & Liu, 2008). An experiment is designed and it occurs in the future. On the other hand, there are many instances where analysis is required of data which have been collected in the past; such an approach can not be an experiment but is known as ex-post-facto research. Ex-post-facto research is very useful in many contexts, such as developing economic models from which forecasts can be made.

In common with any information production, the objective is to support and facilitate decision making. In the case of experimental research, the decision making concerns inference about the relationships investigated. Hence, the variables must be identified, together with appropriate definitions and measures; assumptions should be explicit so that appropriate hypotheses for testing by the experimentation can be formulated expressly. Commonly, such hypotheses concern relationships between independent and dependent variables, to assist examinations of strength and direction of the relationships and, in view of theory and literature, causality. The main stages in experimental design are shown in figure 3.1.

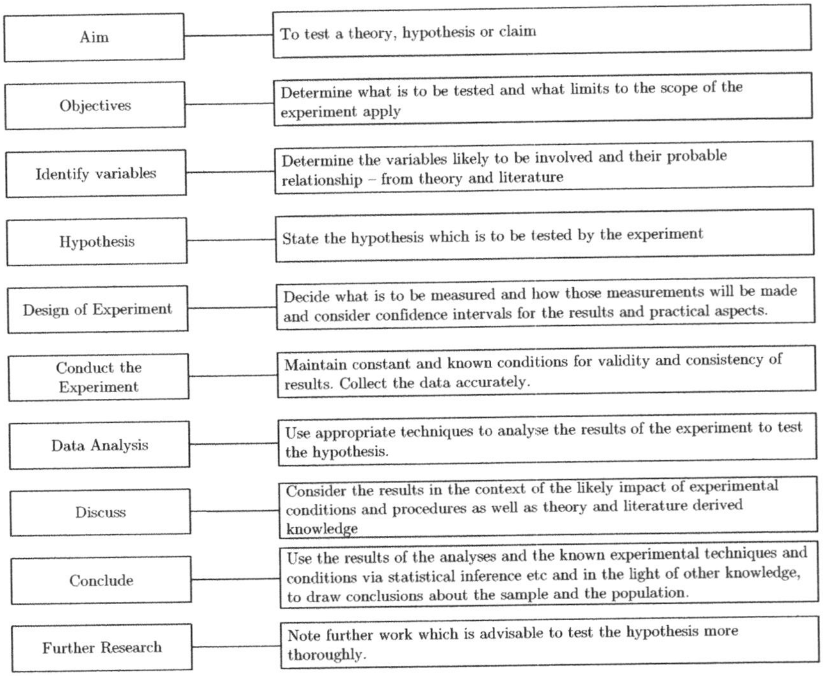

Figure 3.1: Main stages of experimental design (Source: Fellows and Liu (2008))

3.1.2 Motivation

Strictly, empirical studies are those which are concerned with experience gained from experimentation. A debate arises regarding the interpretation of "experience". When it is considered as a natural outcome of human learning process then it incorporates observation, evaluation, memory and recall. Consequently, it can be argued that any one of these four issues can be manipulated or misinterpreted and thus intuitively reliability of human experience can not be trusted. Particularly for decision making practices in construction, employment of personal judgments based upon professional experiences are very common. Hence the production of a unit (a house, an apartment, an hotel) takes much longer time when compared to duration necessary to produce manufactured goods, the construction professionals share a longer time to execute one project. This restricts the number of projects that a construction professional may possibly involve through his life-cycle. The employment of automated machines are recently introduced and yet the overall industry is driven by labor force. According to European Construction Industry Federation, the industry hosts 3,1 million enterprises (EU 27), of which 95% are small and medium size enterprises (SMEs) (FIEC, 2011) On the other hand, each construction work is referred to be unique (Section 1.1). Similarities between projects may be present however, in construction where all variables were strictly tied to each other, variations between two projects would influence dramatic alterations in the outcomes. This is why two identical projects in two distinct locations (such as Italy and Sweden) would alter significantly. Consequently, the odds are unlikely that the past experiences gained from other projects may be applied to the next one and subjective judgments of experiences due to the outlined reasons above shall not be applied to construction. Thereof, an ex-post-facto research is designed.

3.1.3 Modeling

Modeling is the process of constructing a model, a representation of a designed or actual object, process or system, a representation of a reality (Fellows & Liu, 2008). A model must capture and represent the reality being modeled as closely as is practical, it must include the essential features of the reality whilst being reasonably cheap to construct and operate and easy to use. Models may be employed to investigate and/or to predict; for managers predictive models are more valuable, whilst auditing requires investigative models. Particularly, this research aims to develop open-box models. This corresponds to predictive models for which, given all inputs, the outputs may be determined. Irrespective of the type of the model to be developed, all models contain variables which must be identified and quantified for use in the model, together with their inter relationships. Investigation of the literature and limitations gained from time-point of models indicated the appropriate variables to define, isolate and measure, so that performances of the individual variables and their relationships can be evaluated. In validating a model, the model's output resulting from known inputs is compared to realization of the reality, such as an ex-post forecast. The task is performed via a control group of observations randomly sampled from the data set.

3.1.4 Quantitative approach

Essentially, quantitative approaches involve making measurements by collecting data. The approach is built upon previous work which has developed principles, laws and theories to help to decide the data requirements of the particular research project. Two major questions that shall be addressed are: what is to be measured; and how should those measurements be made? The answers are derived from examination of the theory and previous research findings together with the aim and objectives of the research to be conducted; in particular, the hypothesized relationships in the research model. In conducting quantitative research, three main approaches are employed: asking questions of respondents by questionnaires and interviews; carrying out experiments: and 'desk research' using the data collected by others (Fellows & Liu, 2008). Using data collected by others can be very helpful hence it saves considerable amount of time, can be cheap and for studies such a macroeconomics, can be the only viable way of obtaining the data required. Another aspect of experimental design is time dimension, as exemplified in cross-sectional studies, which are observations at one point in time, and longitudinal studies, observations made at multiple time points.

3.2 Linear regression

3.2.1 Overview

Regression analysis is a conceptually simple method for investigating functional relationship among variables (Chatterjee & Hadi, 2006). Principally, a regression equation can be utilized for description and model building, estimation and prediction, control aims (Chatterjee & Hadi, 2006). In accordance with the context of the study, the goal of regression equations developed in this study is set to be estimation and prediction. The variable that is to be predicted is referred to response or dependent variable (Y) whilst determinant(s) which is (are) to be employed to predict the response is called independent or predictor variables ($X_1, X_2, \ldots, X_p$). Accordingly the relationship between Y and the set of X is

$$Y = f(X_1, X_2, \ldots, X_p) + \varepsilon \tag{3.1}$$

where ε is assumed to be a random error representing the discrepancy in the approximation. It accounts for the failure of the model to fit the data exactly. $f(X_1, X_2, \ldots, X_p)$ describes the relationship between Y and $X's$. When the relationship between response and predictors are assumed to be linear the function takes the general form of

$$Y = \beta_0 + \beta_1 X_1 + \beta_2 X_2 + \ldots + \beta_p X_p + \varepsilon \tag{3.2}$$

where $\beta_0, \beta_1, \ldots, \beta_p$ is named as regression coefficients or parameters that are unknown constants to be determined from the data. Regression analysis involve in following steps:

Statement of the problem Regression analysis usually commence with a formulation of the problem that involves in determination of the questions to be addressed by the analysis. The problem statement is referred to be the first and the most important step in regression analysis hence an ill defined problem or a misformulated question may lead to a wasted effort. It can lead to the selection of irrelevant set of variables or to a wrong choice of the statistical method of analysis. Besides, a question that is misformulated may lead to the wrong choice of the model.

Selection of the potentially relevant variables The nest step is to select a set of variables that are thought by the experts in the area of the study to predict the response variable. Literature review, structured or unstructured interviews, and questionnaire surveys are the instruments to achieve the goal.

Data collection Subsequent to the determination of potentially relevant variables, collection of the data under given scope of the study shall be performed. In some fields of research such as medicine, genetics etc. the data are collected in a controlled setting so that factors that are not of primary interest can be held constant. However, in construction economics discipline usually this is not the case. More often, the data are collected under non experimental conditions where very little of none of the external factors can be controlled by the researcher. In either case, the collected data consists of observations on n subjects. Each of these n observations consists of measurements for each of the potentially relevant variables. The data is recorded as in table 2.4.

Table 3.1: Notation for the data used in regression analysis

Observation Number	Response Y	Predictors			
		X_1	X_2	$\ldots$	X_p
1	y_1	x_{11}	x_{12}	$\ldots$	x_{1p}
2	y_2	x_{21}	x_{22}	$\ldots$	x_{2p}
3	y_3	x_{31}	x_{32}	$\ldots$	x_{3p}
$\vdots$	$\vdots$	$\vdots$	$\vdots$	$\vdots$	$\vdots$
n	y_n	x_{n1}	x_{n2}	$\ldots$	x_{np}

Each of the variables in table 3.1 can be identified as either quantitative and qualitative. Within the scope of the problem statement, this study incorporates situations where response variables are classified as quantitative. Therefore, the demonstration of the analysis that is applied when the response is qualitative - logistic regression - is excluded from the manuscript. The data set that is to be utilized in analysis consists of either quantitative and qualitative predictors. In cases, where the response is quantitative and the predictors are both quantitative and qualitative, the analysis is named as *analysis of covariance.*

Model specification Initially, the form of the model that is assumed to be fit to the data set can be proposed by the experts in the area of study based on their knowledge or their objective and/or subjective judgments. The hypothesized model can be either accepted or rejected by the analysis of the data set. It is crucial to select the form of the function expressed in equation 3.1. This function can be classified to two types: *linear* and *non linear* functions. An example of a linear function is given in 3.3.

$$Y = \beta_0 + \beta_1 X_1 + \varepsilon \tag{3.3}$$

while a non linear function is

$$Y = \beta_0 + e^{\beta_1 X_1} + \varepsilon \tag{3.4}$$

It shall be noted that the term linear (non linear) does not indicate the relationship between the response and the predictors. It is associated with the fact that the regression parameters enter the equation linearly. One can see that after natural logarithm transformation equation 3.4 becomes a linear model. Basically, variable transformation is applied in case the variables do not hold normal distribution. It reduces the skewness and induces the symmetry that leads to provide a better fit of the model to the data set as well as assist to fulfill several model assumptions (Tabachnick *et al.*, 2001).

Method of fitting Subsequent to model specification, the researcher deal with the decision of choosing appropriate way of parameter estimation. Models to be developed in the context of this research is limited to bi-variate linear multiple regression models and accordingly *least* *square estimation* (LSE) method will be employed as a default. LSE aims to minimize the sum of squares of the errors. From 3.2, the errors can be written as:

$$\varepsilon_i = y_i - \beta_0 - \beta_1 x_{i1} - \ldots - \beta_p x_{ip},\ i = 1, 2, \ldots, n. \tag{3.5}$$

The sum of squares of these errors is

$$S(\beta_0, \beta_1, \ldots, \beta_p) = \sum_{i=1}^{n} \varepsilon_i^2 = \sum_{i=1}^{n} (y_i - \beta_0 - \beta_1 x_{i1} - \ldots - \beta_p x_{ip})^2. \tag{3.6}$$

It can be shown that the least squares estimates $\hat{\beta}_0, \ldots, \hat{\beta}_p$, which minimize $S(\beta_0, \ldots, \beta_p)$, are given by the solution of a system of linear equations known as the normal equations.

Model fitting The next step in the analysis incorporates estimation of regression parameters or to fit the model to the data set using the chosen estimation method. The estimates of the regression parameters $\beta_0, \beta_1, \ldots, \beta_p$ in (3.2) are denoted by $\hat{\beta}_0, \hat{\beta}_1, \ldots, \hat{\beta}_p$. The estimated equation then becomes:

$$\hat{Y} = \hat{\beta}_0 + \hat{\beta}_1 X_1 + \hat{\beta}_2 X_2 + \ldots + \hat{\beta}_p X_p \tag{3.7}$$

The value $\hat{Y}$ is called the *fitted* value. Using (3.7), one can compute n fitting values, one for each of the n observations in the data set. It is noted that (3.7) can also be utilized to predict the response variable for any given values predictor variables not observed in the data set. In this case the obtained $\hat{Y}$ value is called *predicted* value.

Model criticism and selection The validity of a statistical method, such as regression analysis, depends on various assumptions that are made about the data and the model. The accuracy of the analysis and the conclusions derived from the analysis depends crucially on the validity of these assumptions. Before using (3.7) for any purpose, it is vital to determine whether the specified assumptions are fulfilled. Regression analysis is viewed as an iterative process in which the outputs are used to diagnose, validate, criticize, and possibly modify the inputs. The process shall be repeated until a satisfactory results have been obtained.

3.2.2 The standard regression assumptions

The inferential results that are derived from linear regression analysis are based on summary statistics that are computed from the data. The distribution theory, confidence intervals and tests of hypotheses are valid and have meaning if only the standard regression assumptions are fulfilled. When these assumptions are violated the standard results do not hold and an application of them may lead to serious error. The standard regression assumptions can be outlined as follows:

Assumptions regarding form of the model The model that relates the response Y to the predictors $X_1, X_2, \ldots, X_p$ is assumed to be linear in the regression parameters $\beta_0, \beta_1, \ldots \beta_p$. Accordingly, 3.2 implies that the i^{th}observation can be expressed as:

$$y_i = \beta_0 + \beta_1 x_{i1} + \ldots + \beta_p x_{ip} + \varepsilon_i, \quad i = 1, 2, \ldots, n \tag{3.8}$$

This is referred to as the *linearity* assumption. For simple regression analysis, where there is only one predictor, detection of linearity assumption can be performed by investigation of *scatter-plot*[1] of response and predictor. Checking the linearity in multiple regression is more difficult due to the high dimensionality of the data. When linearity assumption do not hold, transformation of the data may lead to linearity.

Assumptions about the errors The error terms $\varepsilon_1, \varepsilon_2, \ldots, \varepsilon_n$ in 3.8 are assumed to be *independently and identically distributed* (iid) normal random variables each with mean zero and a common variance σ^2. This implies four assumptions:

- The error ε_i, $i = 1, 2, \ldots, n$, has a normal distribution. It is referred to *normality assumption*. The validity of normality assumption can be assessed by examination of appropriate graphs of the residuals.
- The errors ε_i, $i = 1, 2, \ldots, n$ have mean zero
- The errors ε_i, $i = 1, 2, \ldots, n$ have the same (but unknown) variance σ^2. This is the *constant variance* or *homoscedasticity assumption.*
- The errors ε_i, $i = 1, 2, \ldots, n$ are independent of each other (their pairwise covariances are zero). This is referred to be as the *independent - errors assumption.*

Assumptions about the predictors This involves in three assumptions concerning the predictor variables:

- The predictor variables $X_1, X_2, \ldots, X_p$ are non random, that is, the values $x_{1j}, x_{2j}, \ldots x_{nj}$; $j = 1, 2, \ldots, p$ are assumed fixed or selected in advance.
- The values $x_{1j}, x_{2j}, \ldots x_{nj}$; $j = 1, 2, \ldots, p$ are measured without error. This assumption is hardly ever satisfied for construction management discipline hence the observations are not controlled. The presence of error in predictors decreases the accuracy of predictions.
- The predictor variables $X_1, X_2, \ldots, X_p$ are assumed to be linearly independent of each other. When this assumption do not hold, the problem is referred to as the *collinearity* problem.

Assumptions about the observations All observations are equally reliable and have approximately equal role in determining the regression results and in influencing conclusions.

[1]The graph used to explore the potential relationship between a pair of continuous variables *(Minitab, 2010)*

A feature of the method of least squares is that small or minor violations of the underlying assumptions do not invalidate the inferences or conclusions drawn from the analysis dramatically. Gross violations of the model assumptions can, however, seriously distort conclusions.

3.2.3 Detection of violations in model assumptions

Graphical methods play a crucial role in data analysis. They can be regarded as exploratory tools as well as constitute an integral part of confirmatory analysis or statistical inference. Besides, graphical methods can be employed to detect the violations of model assumptions listed in section 3.2.2. The form of a model that represents the relationship between the response and predictor variables should be based on the theoretical background or the hypothesis to be tested. However, if no prior information about the form of the model is available, the data may be employed to suggest the model. Examination of interrelationships and trends between variables may explore the form of mathematical expression.

Data analysis commences with the examination of each variable that retains in the analysis. The aim is to have a general idea regarding distribution of each individual variable. Usually, histograms[2] or box plots[3] are employed for this purpose. They indicate whether the variable is skewed or symmetric. If the former case is present, variable transformation shall be applied. For a high positively skewed variable *logarithmic transformation* is suggested (Chatterjee & Hadi, 2006). One dimensional graphs can also be utilized to detect the influential observations with extreme values (outliers) in the data set. However, no observation shall be deleted at this stage of analysis.

Subsequently, two or multi-dimensional graphs such as scatter plots are employed to examine interrelationship between response and predictors. In case where there are more than one predictor matrix scatter-plot is useful to investigate the patterns. In consistent with the shape of the relationship that is identified in scatter-plot, equation 3.1can be defined, hypothetically.

The graphs mentioned up to now are useful in data checking and the model formulation steps. The graphs after fitting a model to the data help in checking the assumptions and in assessing the adequacy of the fit of a given model. These graphs incorporates in investigation of residuals[4]. The following plots of residuals can be utilized to verify the linearity and normality assumptions:

1. *Normal probability plot of the standardized residuals:* This is a plot of the ordered standardized residuals versus so called normal scores. The normal scores are what one would expect to obtain when taking a sample size of n from a standard normal distribution. In case the residuals are distributed normally, the ordered residuals should be approximately the same as the ordered normal scores. Under normality assumption, this plot should resemble a (nearly) straight line with an intercept of zero and a slope of one.

2. *Scatter plots of the standardized residual against each of the predictor variables:* Under the standard assumptions, the standardized residuals are uncorrelated with each of the predictor variables. If the assumptions hold, this plot should be random scatter of points. Any discernible pattern in this plot may indicate violation of some assumptions.

[2] A graph used to assess the shape and spread of continuous sample data (Minitab, 2010)

[3] A graphical summary of the distribution of a sample that shows its shape, central tendency, and variability (Minitab, 2010).

[4] The difference between an observed value (Y) and its corresponding fitted value ($\hat{Y}$).

3. *Scatter plot of the standardized residual versus the fitted values:* Under the standard assumptions, the standardized residuals are also uncorrelated with the fitted values and therefore, this plot must also be a random scatter of points.

4. *Index plot of the standardized residuals:* This is the diagnostic plot which is employed to display the standardized residuals versus the observation number. Hence, the observations employed in this study were executed in the past and the data is collected in between a long period of time, index plot is vital to check the assumption of independence of the error. One should expect to observe the points scattered randomly within a horizontal band around zero.

3.3 Artificial Neural Networks

3.3.1 Overview

ANNs are computational models that are developed to mimic basic biological neural systems - the human brain particularly (Zhang *et al.*, 1998). Recent researches have shown that ANNs have powerful pattern classification and recognition capabilities. They are able to learn from and generalize from experience. Widrow *et al.* (1994) reported wide variety of ANNs applications in fields such as business, industry and science.

Sharda (1994) indicated forecasting as one of the major application area of ANNs. There are unique features of ANNs that enable usage of them as powerful forecasting tools. First, ANNs are data-driven self adaptive methods and do not require priori assumptions about the models for issues under investigation (Zhang *et al.*, 1998). They learn from the observations in the data and try to minimize the error of predictions by correcting the weights of relations in each iteration. Zhang *et al.* (1998) argues that generally it is easier to find data than to reach a theoretical knowledge regarding underlying principles governing the systems from which data is generated. Second, ANNs can generalization capabilities. This is to say, subsequent to learning from a sample ANNs can generalize the pattern even for the unseen part of the population. Third, results of Irie & Miyake (1988); Hornik *et al.* (1989) rigorously show that ANNs are universal functional approximators and a network can approximate any continuous function to any desired accuracy. Last but not least, ANNs are capable of explain non-linear relations as well as interactions between variables.

Previously, various types of ANNs have been developed such as radial basis functional networks (Park & Sandberg, 1991), ridge polynomial networks (Shin & Ghosh, 1995), and wavelet networks (Zhang & Benveniste, 1992). Yet, in the following sections we focus on multi layer perceptrons (MLP) and general regression neural network (GRN).

3.3.2 Multi layer perceptrons

An MLP is typically composed of several layers of nodes (Figure 3.2). The first set of nodes is fed by the input variables (vector) and named as input buffer or layer. The input layer is followed by one or more hidden layers. Optimum number of neurons in the hidden layer(s) remains as one of the biggest debate in ANNs research. Empirical evidences from several studies suggest different rules yet, there is no consensus on this debate. The outcome layer is composed

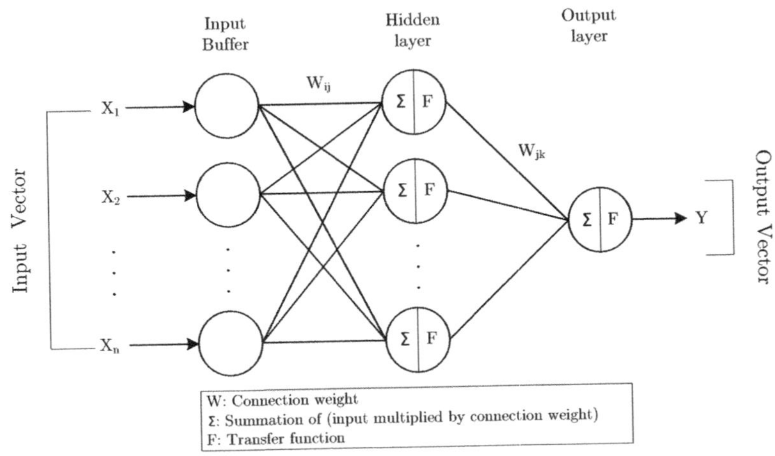

Figure 3.2: A typical feedforward network (MLP)

of one or more neurons. This is to say MLP can be utilized to predict more than one outcome at a time. Following sections are allocated to the issues related to MLP. Related parameters have to be optimized specifically for the problem under investigation since there is no clear cut theory works on every case.

3.3.2.1 The network architecture

As indicated, an MLP is composed of layers of nodes. While designing an MLP one must determine followings:

The number of hidden layers and nodes Previous applications of neural networks remarked that the number of hidden layers and nodes plays a crucial role for the success of the results. The motivation stands for it, is finding patterns and generalizing non linear relations between input and output variables is performed with the help of neurons in the hidden layer(s). Zhang *et al.* (1998) stated that lacking of hidden neurons an MLP shows of no difference than a linear statistical forecasting models. Studies by others (Cybenko, 1989; Hornik *et al.*, 1989)show that MLP with one hidden layer is sufficient for approximate any complex nonlinear function with any desired accuracy. Particularly, in construction economics, all developed ANNs models (Chapter 2) are composed of one hidden layer. This concretes the idea of designing MLP with one hidden layer in this study.

Number of nodes in the hidden layer always receives higher attention and more complicated. Generally, smaller number of hidden neurons is preferred due to better generalization ability and avoiding over fitting. Yet, when the number of hidden neurons is too small than MLP performs under its capacity and can not capture the unique properties of noise data. There is no theoretical basis for selecting this parameter. Most common way in determining the number hidden nodes is via experiments or by trial and error (Zhang *et al.*, 1998).

For construction cost estimation problems Hegazy *et al.* (1994) suggest "$0.75n$", "n", "$2n+1$" where n is the number of input variables (nodes). Different than those, in other fields of applications "$2n$" (Wong, 1991), "$n/2$" (Kang, 1991) are also offered. However, none of these heuristic choices works well for all problems (Zhang *et al.*, 1998).

The number of input nodes The number of input nodes corresponds to the number of predictor variables to be employed to predict the desired outcome.The selection of this parameter shall be considered as development of the model, variable selection procedure. Number of input nodes shall be optimized with attention hence it directly influences the learning and prediction capability of an MLP.

In construction cost estimation applications of MLP, the number of input nodes is determined subsequent to regression analysis (Emsley *et al.*, 2002; Sonmez, 2004; Karanci, 2010). The significant variables of regression models are later employed as inputs of the MLP models. Recall that MLP has capacity to explain nonlinear relations and interactions. For comparison aims, some [e.g. (Emsley *et al.*, 2002)] develop MLP models by using all available predictors as inputs. Last but not least, Sato *et al.* (2005); Shastri *et al.* (1998) advocate eliminating unimportant factors to establish parsimonious models which in turn increase generalization capacity of ANNs.

In our study, similar to the applications in construction cost estimations, we employ significant regression variables as inputs of MLP models to be developed.

The number of output nodes The number of output nodes is generally easy to specify and based on the problem under investigation. Yet, the aims of this research requires a higher attention to the selection of this parameter as well. This study is not the first one that offers an iterative approach to a forecasting problem. Zhang *et al.* (1998) indicated two types of forecasting: one-step-ahead (which uses one output node) and multi-step-forecasting. Two ways of multi-step-forecasts are reported in the literature. The first is the direct approach is to let the MLP have several output nodes to directly forecast each step into the future. The second is the iterative one (the approach adapted in this research), in which the forecast values are iteratively used as inputs for the next forecasts (Recall Figure 1.4 and 1.5). In this approach, only one output is node is necessary. Results of Weigend *et al.* (1992) demonstrate iterated single step predictions performed better than direct multi step ones. Parallel, Hill *et al.* (1994) conclude iterated approach is superior to direct approach. However, results by Zhang (1994) favor direct prediction is much better than the iterated method.

3.3.2.2 Activation function

Activation function, so called transfer function, determines the relationship between inputs and outputs of a node and a network. The definition indicates that activation function is crucial hence it introduces the degree of non linearity which is the primary reason to employ an MLP. In practice, 4 types of activation function is widely used:

1. The sigmoid (logistic) function:

$$f(x) = \frac{1}{1+\exp(-x)}$$

2. The hyperbolic tangent (tanh) function:

$$f(x) = \frac{[\exp(x) - \exp(-x)]}{[\exp(x) + \exp(-x)]}$$

3. The sine or cosine function:

$$f(x) = \sin(x)\ ;\ f(x) = \cos(x)$$

4. The linear function:

$$f(x) = x$$

Emsley *et al.* (2002) advocates using hyperbolic tangent function provides superior results for cost estimation, given the fact that tanh function approximately linear at zero, requiring the variables to be normalized and the weights set to very small values. This indicates curve areas in the function is employed to model non linear behavior subsequent to assumed linear behavior at the commence of training.

3.3.2.3 Training algorithm

Training algorithm is the part where computational optimization takes part. According to Zhang *et al.* (1998):

> *"A neural network training is an unconstrained nonlinear minimization problem in which arc weights of a network are iteratively modified to minimize the overall mean or total squared error between the desired and actual output values for all output nodes over all input patterns."*

"Back-propagation" was the first successful algorithm that minimizes the arch weights in MLP. Thereof, it has a long tradition and been widely employed in every application field including construction cost estimation. Yet, last two decades researchers tend to favor other algorithms as faster and more likely to ind the global optimum. Particularly in this research, "Conjugate Gradient Descent" method, belonging to the category of "second order" optimization methods, is selected and employed to train the network models.

3.3.2.4 Data normalization

ANNs uses only numeric values as inputs. Basically, requirement to normalize the data is rely on meeting training algorithms requirements. The inputs are normalized the values between (-1,1) or (0,1) and later the output node is transformed back to the range of actual outputs.

For the aims of this research linear transformation offered by Lapedes & Farber (1988) is applied. The transformation resulted the values between (0,1). Accordingly the transformation function is given as:

$$x_n = \frac{(x_0 - x_{min})}{(x_{max} - x_{min})}$$

where x_n and x_0 represent the normalized and original data; x_{min} and x_{max} are the minimum and maximum along the columns and rows, respectively. It is also noted that linear transformation function is applied in construction economics by Hegazy & Ayed (1998).

3.3.2.5 Training and test sample

As mentioned earlier, the data set shall be divided into two subgroups namely as training and test samples. Training sample is employed to train the MLP regarding the general patterns in the data set. Test sample is used by the trained MLP to predict expected values. Later, predicted and actual values of test sample is to be compared in terms of performance measures.

There is no specific rule for the sample sizes of training and test groups. One important factor to be considered is the representative ability of both training and test samples over selected population. Applications of ANNs indicated several heuristic rules. Most authors select 90% vs. 10%, 80% vs. 20% or 70% vs. 30%, etc. Details more on training and testing sample can be found in Chapter 4.

3.3.2.6 Performance measures

The ultimate and most important performance measurement for any ANNs is its prediction accuracy that can be reached beyond the training data. There are a number of measures of accuracy in the forecasting literature and each has advantages and limitations. The most frequently used are:

- mean absolute deviation (MAD) $= \frac{\sum |e_t|}{N}$;
- the sum of squared error (SSE) $= \sum (e_t)^2$;
- the mean squared error (MSE) $= \frac{\sum (e_t)^2}{N}$;
- the root mean squared error (RMSE) $= \sqrt{MSE}$;
- the mean absolute percentage error (MAPE) $= \frac{1}{N} \sum |\frac{e_t}{y_t}|(100)$,

where e_t is the individual forecast error; y_t is the actual value and N is the number of error terms. In our research, MAPE and RMSE are taken as measures of performance.

3.3.3 General regression neural network

A general regression neural (GRN) network is a memory-based network that provides estimates of continuous variables and converges to the underlying (linear or nonlinear) regression surface (Specht, 1991). This general regression neural network is a one-pass learning algorithm with a highly parallel structure. Even with sparse data in a multidimensional measurement space, the algorithm provides smooth transitions from one observed value to another. The algorithmic form can be used for any regression problem in which an assumption of linearity is not justified.

The parallel network form should find use in applications such as learning the dynamics of a plant model for prediction or control.

A typical GRN Net block diagram is presented in Figure 3.3. The Pattern Layer contains one node for each training case. Presenting a training case to the net consists here of presenting two independent numeric values. Each neuron in the pattern layer computes its distance from the presented case. The values passed to the Numerator and Denominator Nodes are functions of the distance and the dependent value.

The two nodes in the Summation Layer sum its inputs, while the Output Node divides them to generate the prediction. The distance function computed in the Pattern Layer neurons uses "smoothing factors"; every input has its own "smoothing factor" value. With a single input, the greater the value of the smoothing factor, the more significant distant training cases become for the predicted value. With 2 inputs, the smoothing factor relates to the distance along one axis on a plane, and in general, with multiple inputs, to one dimension in multi-dimensional space.

GRN Architecture Training a GRN net consists of optimizing smoothing factors to minimize the error on the training set, and the conjugate gradient descent optimization method is used to accomplish that. The error measure used during training to evaluate different sets of smoothing factors is the mean squared error. However, when computing the squared error for a training case, that case is temporarily excluded from the Pattern Layer. This is because the excluded neuron would compute a zero distance, making other neurons insignificant in the computation of the prediction.

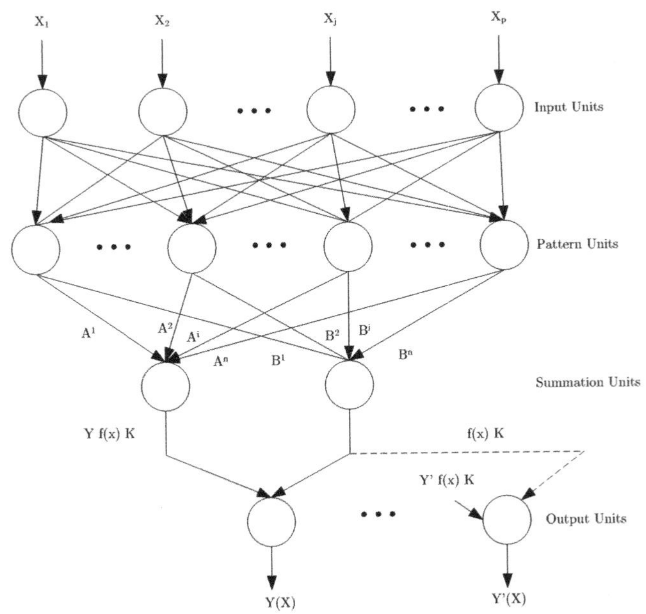

Figure 3.3: GRN Net block diagram (Specht, 1991)

Lastly, two clear advantages of GRN Net in comparison to MLP are: (1)they train fast and (2)they do not require any topology specification.

3.4 Concluding Remarks

This chapter can be regarded as big parentheses in the manuscript that addresses those who are lacking information on research methodology and method of analysis. These methods are employed for testing and development of hypotheses and later to compare the results of the models.

To sum up, quantitative methods are employed in this research. Hence, the research team does not have any control on the historical project information a particular branch of empirical study, an ex post facto research, is designed. In order to model desired outputs two techniques of analysis that have been widely employed in the literature were selected. First, linear regression which assumes a linear relationship between predictor and response variables is presented. Shortcomings of the model and necessity to fulfill model assumptions is underlined. Second, artificial neural networks, computational instruments that mimic learning principles of human brain, is showed. Two alternatives of neural networks were considered in this study, namely multi layer perceptrons and general regression neural networks. In multi layer perceptrons, architecture of a network must be determined after heuristic processes that were underlined. The capabilities of capturing non linear relationships as well as interactions between inputs and output of the model were mentioned. Performance measurements that are to be utilized to compare the results of the models were given. General regression neural net, on the other hand, is one-pass learning algorithm with a highly parallel structure. Advantages of GRN Net in comparision to MLP are: (1)they train fast and (2)they do not require any topology specification.

4 Sample

So far, reported chapters were associated with contextual background of the research. From this chapter on, the main concentration is set to outline the execution phase of the study. This commences with exhaustive demonstration of properties of the sample and modeling procedure.

Like in any empirical work, sample plays a crucial role for reliability of the results. It shall represent properties of the selected population within the content of the research to the same extent. This chapter commences with general description of the data. This leads to discussion regarding representative ability of the sample. It continues with summarizing restrictions applied to the sample due to the scope and presentation of available variables. Later, descriptive statistics are given.

4.1 Overview

The data employed in this research was gathered from Cost Information Centre of German Chamber of Architects GmbH (BKI). BKI has a traditional reputation in providing reliable and accurate building cost information to the German practitioners for the last five decades. Using secondary data from a professional company provided a much higher number of study units than could be reached using primary data, as well as a presumed increase in the reliability of relevant project information.

Initially, the data set was composed of 1858 building information. Yet, observations requires to be filtered prior to involve in the analysis. Following filters were applied on data:

- *Type of construction works*: The full data set consisted of renovation, refurbishment, new construction, expansion to or conversion of existing buildings, type of construction works. The research scope was limited to only new construction works, thereof the rest was excluded.

- *Type of product*: Full data set was composed of two general product types, namely buildings (offices, residences, etc.) and landscape works (parks, etc.). Parallel to the scope landscape works were excluded.

- *Level of information (missing values)*: Please recall Figure 1.4, 1.5, and Table 1.1. Accordingly, involving in the analysis, information level of an observation had to be detailed at least to the second level of DIN 276.

- *Measurement mistakes*: Careful investigation of the data set was performed to prevent measurement mistakes of retained observations. Some project information in the data violated common sense such as negatively signed construction duration. These were marked as measurement mistakes and excluded.

- *Setting range of scope*: Logical ranges were set for the project size. In this context, projects that are smaller than 200 m^2 in gross external floor area and 200.000 € in cost of structure, construction works were excluded. Due to observed limitation in scope of works, these projects did not represent the true figure in the market.

Table 4.1: Qualitative attributes of the sample

Qualitative attribute	Number	Percentage
Facility type (FAC)		
Residential	301	46%
Office	65	10%
Industrial	147	22%
Educational	72	11%
Other	72	11%
Ground conditions (GC)		
Soft	154	23%
Moderate	487	74%
Hard	16	2%
Site access (SA)		
Free	525	80%
Semi restricted	86	13%
Restricted	46	7%
Topography (TOP)		
Flat	421	64%
Gentle sloped	163	25%
Steep	73	11%
Market conditions (MC)		
Stagnation	193	29%
Stable	339	52%
Boost	125	19%
Client type (CT)		
Public	166	25%
Private	491	75%

Note: Sample size "N" is equal to 657.

Defined filters exclude approximately 65% of the data set and the analysis were conducted with 657 remaining projects. For the accuracy of a forecast model, homogeneity of the data is at paramount. Otherwise, the effect of additional variance within the data set might effect the estimation. Despite of strict filtering, the sample size was the largest when compared to relevant studies in the same field. This underlines significance of the study. Yet, it shall be emphasized that it contained large variety of facility types from residential to cultural that shows distinct structural properties.

The sample consists of nine facility types where residential developments were dominant compared to others. Parallel to the pattern obtained from the sample, similar dominant figure was also observed for the population. According to German Federal Statistical Office (2012), residential construction forms approximately half of the new construction work in Germany, annually. Table 4.1 showed qualitative attributes, levels (so called categories), and percentages in the sample.One could see in Table 4.1 that site access to majority of the sampled projects were free in all direction. Common sense suggested this is generally the case when development is located out of downtown. Historically, German cities have expanded horizontally which created a homogenous population density. Only population of Berlin exceeds 3 million inhabitants; followed by Hamburg, Munich and Cologne whose population are just above 1 million (DESTATIS, 2012). Even cities like Frankfurt, the finance capital of European Union; Stuttgart, the hometown of world manufacturing giants such as Daimler and Bosch, are below 700.000 inhabitants (DESTATIS, 2012). The properties of the population and the sample presented similarities as it was shown. This indicated the sample has capacity to represent selected population.

All sampled projects were executed between 1973 and 2009. This indicated a large span of time. To eliminate the effect of time on the capital all cost related variables are adjusted to 3. quarter of 2011 by using German construction cost price index published by German Statistical Office, quarterly. All cost information is VAT included. All data points are actual values that is incurred in the construction site.

Table 4.2: Comparison between training and test samples

Facility type	Training Sample		Test Sample	
	No. obs.	Perct.	No. obs.	Perct.
Residential	281	47%	20	35%
Educational	65	11%	7	12%
Industrial	132	22%	15	26%
Office	55	9%	10	18%
Other	67	11%	5	9%
Sum	*600*	*100%*	*57*	*100%*

As indicated in Chapter 3, observations must be divided into two sub samples: One to conduct modeling, so called training sample, and one to measure prediction performance of the models, namely test sample. 57 out of 657 observations were selected randomly. These observations composed the test sample. Remaining 600 building project information was utilized to recognize the patterns in the data and therefore compose training sample. The ratio of test to training sample is approximately 10% to 90%. In addition, the number of observation in the test sample is adequate to draw unbiased statistical inferences and to conduct formal test of hypothesis for comparison of the results. Last but not least, test sample shall be representative of training sample so that developed models will be tested under similar conditions. One way to verify this is to compare percentages of facility types in both samples. This was given in Table 4.2.

4.2 Restrictions and Predictor Variables

Although available data set presents a variety of detailed building information in accordance with relevant German norms, time-point of models to be developed substantially limits predictor variables that can be retained in the data set. It was intended to have an objective opinion regarding the predictors that are available to an estimator through establishing the basis of a project, in practice.

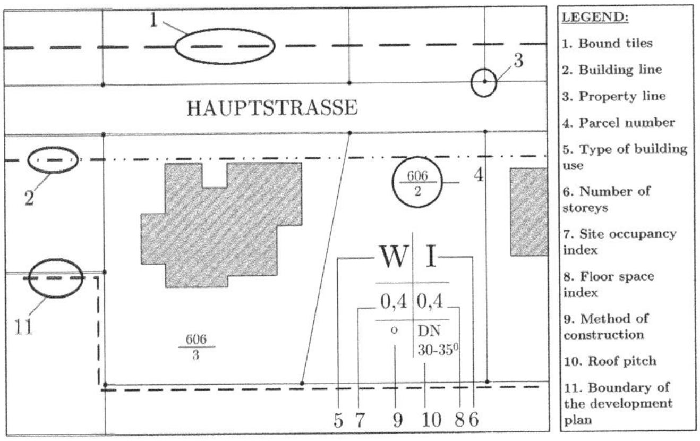

Figure 4.1: An example of a development plan issued by Stadtplanungsamt

Properties of a building to be constructed on a particular plot are strictly regulated by the German municipal institutions called as *Stadtplanungsamt* or *Abteilung für Stadtplanung*. Figure 4.1 illustrates an example of a development plan of a particular plot in Germany. In Figure 4.1, some key macro determinants of a building is already revealed to an architect. Accordingly, (5) indicates the type of facility, "W" stands for "wohnung (residential)"; (6) is the number of full storeys above ground; (7) is, on the other hand, the ratio of footprint area of building to the overall plot.

In addition, qualitative properties such as topography of the plot, ground conditions and ease of access to construction site can be tracked prior to commencement of the design (see Table 4.1). In conclusion, it was assumed that numerical variables that are listed in Table 4.3 can be available to an architect at the conceptual design phase.

Table 4.3: Available quantitative predictor variables to an estimator at the basic evaluation stage of HOAI

Var.	Description	Type	Mean	St. Dev.	Median
GEFA	Gross external floor area (m^2)	CN	2,605	3,617	1,220
GBV	Gross building volume (m^3)	CN	10,521	16,479	4,221
NoS	Number of storeys	DN	3.34	1.80	3
NoS_{ag}	Number of storeys above ground	DN	2.63	1.53	2
NoS_{bg}	Number of storeys below ground	DN	0.71	0.56	1
PA	Plot area (m^2)	DN	4,773	8,289	2,157
DPA	Developed plot area (m^2)	CN	1,139	1,723	579
ASH	Average storey height (m)	CN	3.78	1.34	3.37
AFS	Average floor size (m^2)	CN	780	942	418
LI	BKI Location index	CN	1.06	0.1	1.04

Note: "CN" and "DN" stand for continuous numeric, and discrete numeric variables, respectively. Sample size is equal to 657 and no observation has missing value.

4.3 Response Variables

Eventually, the data set is composed of not only predictor (independent), but also response (dependent) variables. The response variables are reference quantities of second level cost groups of construction works (recall Table 1.1), second and first level construction cost, and construction duration.

Table 4.4: Response variables and their descriptive statistics

Var.	Description	Mean	St Dev	Median
m310	Excavation volume (m^3)	3,051	5,332	1,066
m320	Foundation area (m^2)	1,093	1,692	555
m330	External walls area (m^2)	1,543	1,840	831
m340	Internal walls area (m^2)	1,828	2,801	790
m350	Ceilings and floors area (m^2)	1,410	2,272	493
m360	Roof area (m^2)	1,227	1,785	650
k310	Cost of excavation (€)	91,999	308,306	21,268
k320	Cost of foundation (€)	254,720	399,578	111,692
k330	Cost of external walls (€)	680,847	1,106,490	296,830
k340	Cost of internal walls (€)	451,806	880,938	173,365
k350	Cost of ceilings and floorings (€)	413,495	741,325	132,446
k360	Cost of roof (€)	372,545	565,502	172,896
k370	Cost structural fitments (€)	88,714	254,715	5,686
k390	Cost of other construction related act.(€)	107,960	230,144	28,771
k410	Cost of sewerage water and gas sys. (€)	159,220	322,321	65,293
k420	Cost of heat supply systems (€)	164,434	294,752	71,582
k430	Cost of air treatment systems (€)	151,256	492,574	10,401
k440	Cost of power installations (€)	196,425	447,700	62,051
k450	Cost of telecommunications (€)	50,560	136,230	6,128
k460	Cost of transport systems (€)	37,185	97,250	0
k470	Cost of function related equip. & fit. (€)	172,486	1,671,051	0
k480	Cost of building automation (€)	141,78	117,280	0
k490	Cost of other services related work (€)	1,821	19,383	0
k300	Cost of structure - construction works (€)	2,475,406	3,964,477	1,042,493
k400	Cost of structure - services (€)	947,566	2,634,594	259,852
k300+k400	Cost of structure (€)	3,422,972	6,295,445	1,325,462
CD	Construction duration (calendar days)	459	273	366

Note: All costs are adjusted to 3. quarter 2011 prices

Discussion regarding representation of output variables has been addressed by Emsley *et al.* (2002). The readers are referred to Section 2.1.2 for detailed information. Initially, it had been intended to adopt indicators as response variables offered by Emsley *et al.* (2002). Yet, results of the pilot analysis revealed that explaining variance of the indicators would be arduous due to lacking information. Accordingly, we rejected the use of indicator variables as outputs in our models. Instead, it was intended to explain the variation in the value of interest. Table 4.4 listed all output variables to be employed and their corresponding descriptive statistics.

In Table 4.4 no variables has missing values. Yet, one shall distinguish the difference between missing values and "*zero*"s. If one building has no particular excavation work than cost of excavation and excavation volume must be equal to zero. Some variables, such as transport systems, has median value of zero. This indicates that at least half of the observations do not consist of transport systems works. Understanding this is crucial hence observations that has value of zero will not be included in the modeling or testing phase. It was assumed that an architect is capable of judging the necessity to conduct a specific task when first approach to a client is initiated.

4.4 Supporting the Hypotheses - Pilot Analysis

Subsequent to exhaustive demonstration of available predictors and target values (responses) to be estimated, convincing readers regarding credibility of selected method of analysis can be accomplished by illustrating bi variate scatter plots of selected variables.

Recall that main test of hypothesis in this study relies on the Frameworks that were demonstrated in detail in Section 1.3. Frameworks, which adopted multi-way approach (#1, #2, #5) was constructed based on the findings of the pilot study. The pilot analysis of the data set reveals a significant linear dependency between reference unit quantities, such as external walls area, and predictor variables that are available through establishing the basis of the project, such as gross external floor area. Strength of this linear relationship can best be presented in bi variate scatterplot that is shown in Figure 4.2. It is evident that linear functions can be employed to explain the variance in the values of interests, in this case reference unit quantities of second level cost groups of structure.

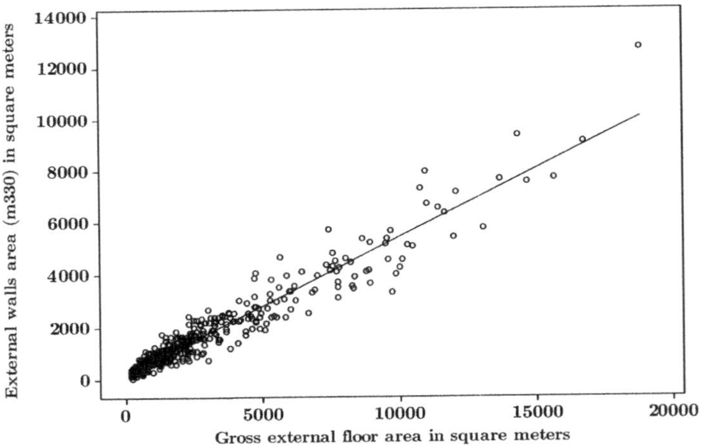

Figure 4.2: Scatterplot of external walls area and gross external floor area

In multi-way approach, it was assumed that estimated values of reference unit quantities can be used as additional inputs to predict second and first level cost groups of structure. In Chapter 2, linear models along with artificial neural networks were proposed to conduct the task as the appropriate method of analysis. This assumption again relies on the pilot analysis of the data set, too. Demonstrating credibility of form of the relationship, a simple linear regression model aiming to predict external walls area using gross external floor area as the independent variable was developed. Predicted (fitted) values of external walls area, resulted from this relatively simple model, was than shown in relationship with actual values of cost of external walls in Figure 4.3. One can see that a linear function can be employed to account for the variance in the actual cost of external walls, after natural logarithm transformation is applied on both axes.

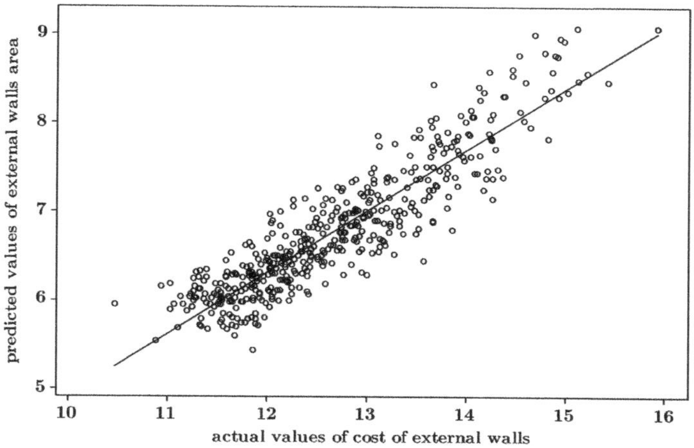

Figure 4.3: Scatterplot between predicted values of external walls area and actual values of cost of external walls

In conclusion, it is evident with the findings of the pilot analysis that was shown in Figure 4.2, linear relationships are present between reference unit quantities of second level cost groups and macro project determinants which are available through establishing the basis of a project. In addition, Figure 4.3 demonstrated clearly that fitted values resulted by even relatively simple models show a significant linear relationship with the actual values of second level cost groups of structure. These relationships led us to develop and offer Frameworks which are unique to this study and presented exhaustively in Section 1.3.

4.5 Concluding Remarks

Representative ability of the sample over selected population plays a vital role in applicability of results in practice. It was argued that attributes of the sample and selected population show significant similarities. Size and quality of the sample matters for the plausibility of the models. The size of the sample is large enough to draw adequate statistical inferences. Whereas, quality of the measurements were not under the control of the research team. Yet, reputation of the supporting company as a data provider, as well as long tradition of German standardization and definitions on construction cost data enhances confidence on the quality of the measurements.

Restrictions due to the time-point of prediction models, limited number of available predictors to an estimator. Categorical and numerical predictors were listed in Table 4.1 and 4.3, respectively. Reverse to raised discussion regarding representation of the response variables by others, output variables are formed in their raw forms. It was noted that high level of unexplained variance remains when indicators are employed as outputs. Summary statistics and descriptions of output variables can be tracked in Table 4.4.

5 Results of Analysis

Chapter 5, results of analysis, presents exhaustive demonstration of the models developed to estimate reference unit quantities, construction costs at different hierarchical levels, and construction duration. In addition, general discussion regarding developed models and basic comparison in between alternative method of analysis, artificial neural networks and linear regression were conducted and reported.

All descriptive and inferential statistical analysis were conducted with the help of a commercial software package: "Minitab V.16.0". Minitab provides flexible and user friendly interface along with powerful graphical engine. In addition, compatibility of Minitab with Microsoft Excel, in which the data used in this study has been stored, is flawless. What is more to the point, computation capabilities of software is sufficient enough to conduct the necessary statistical procedures that are conducted in this study. These procedures incorporates with descriptive statistics (Chapter 4), stepwise and general regression routine (Chapter 5 and Chapter 7), and analysis of variance (Chapter 6). Alternative method of analysis, artificial neural network, on the other hand, is conducted with another commercial Excel Add-In, Palisade Decision Suite - Neural Tools. As an Excel Add-In, Neural Tools are compatible with the data, presented significant amount of ease of use. In addition, with help of its advanced optimisation algorithm, time to train multi-layer perceptron neural network models significantly reduced when compared to its counterparts.

Training sample was employed through modeling procedure. Categorical variables of ground conditions, topography, site access, and market conditions, were transformed into dummy numerical values. This enabled to use the sample as a whole instead of small sub-samples, enhanced number of observations and therefore capacity to generalize. Overall, 6 models for reference unit quantities, 17 models for second level cost groups of structure, 4 models for first level cost groups of structure, 2 model for total cost of structure and 2 models for construction duration were developed for each method of analysis, namely linear regression and artificial neural networks.

Modeling procedure would be conducted in accordance with the frameworks of the research as illustrated in Figure 1.4 and 1.5. Framework #1, #2, and #5 corresponded to multi-way forecasting. This required an iterative process where predicted values of reference unit quantities would be fed back to the analysis to predict second target values: construction costs and duration. Framework #3, #4 and #6, on the other hand, indicated one-step ahead forecasting. Predictors were limited to those that are available at the HOAI - basic evaluation stage (see Table 4.3 and 4.1). First method of analysis, which was employed to predict future values of the response variables, was linear regression, as indicated in Chapter 3. Following artificial neural networks models were developed relying on the outcomes of the linear regression models. The outcomes of analysis was demonstrated with respect to the value of interest: Reference unit quantities, construction costs, and construction duration.

5.1 Reference Unit Quantities

Reference unit quantities of second level cost groups were required to be employed in Frameworks #1, #2, and #5. This involved in estimating quantities of building elements according to German norms. It was hypothesized that, this, in turn, may decrease the level of uncertainty to an extent which resulted in increased prediction accuracy for cost of structure and construction duration.

In order to obtain an initial opinion regarding modeling of reference unit quantities, structured interviews with German practitioners were conducted. This resulted in obtaining initial (assumed) model determinants (Table 5.1). Later, stepwise regression (forward and backward elimination) was employed to test significance level of hypothesized numerical variables. Predictors of which significance level according to *t-test* were above 10% were excluded from the models. At last, assumed categorical variables were added to the model. F-test was employed to test whether categorization of observations with respect to assumed qualitative factors provided higher explained variance in response variables. This formed the final model. Subsequent to determination of significant factors, outlier elimination was performed attentively. Last but not least, the parameters of the equation were estimated by ordinary least square estimation method.

Final model was checked if any of linear regression model assumptions were violated. Residual plots were used for this aim. It was evident that all developed models were violated couple of model assumptions. First, error terms were not normally distributed. Second, variances of error terms were heteroscedastic. This is to say, variances of error depended on the value of response variable. Consequently, variable transformation to responses were applied to fulfill the assumptions of normality and heteroscedasticity of error terms. Variable transformation was performed with an automated procedure by Minitab, namely Box-Cox transformation. With an iterative process, Box-Cox helps to determine the best form of response variable of interest to fulfill model assumptions of linear regression models that were outlined in Subsection 3.2.2.

Due to the nature of construction and predictor variables, multi-collinearity exists. This was not an experimental study where the study units could be isolated. Nearly, all predictors were correlated to each other to an extent. However, the research showed that the effect of multi-collinearity on this study's primary target, prediction, is none whatsoever (Chatterjee & Hadi, 2006). Thus, collinearity was not addressed as an issue through development of the *predictive* models. Yet, due to collinearity, parameters of developed models were interpreted attentively.

Table 5.1 outlined candidate predictors that were included in the initial models. According to the results of forward and backward stepwise regression at 10% significance level, along with expert opinion, gross external floor area (GEFA), gross building volume (GBV), facility type (FAC) and average floor size (AFS) appeared in all initial models. These variables were followed by developed plot area (DPA) and average storey height (ASH), which were included all initial models except excavation volume (m310). Models apart from external walls (m330) and floor and ceilings (m350) contained number of storeys above ground level (NoS_{ag}) initially. The

Table 5.1: Candidate predictors for reference unit quantity models

Var.	Description	Initial model in which the factor is included					
		m310	m320	m330	m340	m350	m360
GEFA	Gross external floor area	■	■	■	■	■	■
GBV	Gross building volume	■	■	■	■	■	■
NoS	Number of storeys			■			
NoS_{ag}	NoS above ground	■	■		■		■
NoS_{bg}	NoS below ground	■			■	■	
PA	Plot area			■	■	■	
DPA	Developed plot area		■	■	■	■	■
ASH	Average storey height		■	■	■	■	■
AFS	Average floor size	■	■	■	■	■	■
LI	BKI Location index	■					
FAC	Facility type	■	■	■	■	■	■
GC	Ground conditions		■	■		■	
SA	Site access				■		
TOP	Topography	■		■		■	

Note: Market conditions and client type were not hypothesized by the experts and therefore are not listed in the table.

frequency of appearance for the candidate predictors could be investigated in detail in Table 5.1.

The results of ordinary least square parameter estimation aided to test hypotheses of whether initially included variables made significant contribution to the explained variance in the model responses. This was controlled by probability of t statistics at 5% significance level. Estimated parameters of significant predictors for reference unit quantity models were presented in Table 5.2.

Interpretation of regression coefficients is source of confusion for multiple linear regression models. The multiple regression equation does not represent a line but a *hyperplane*. Common exercise is to assume holding all predictors constant but one, and vary its value to observe the departure from the original fit of the response. While this holds valid in experimental studies this is absolutely not the case for observational research. Particularly, in our practice, construction, the predictor variables were inherently related and holding some of them constant while varying the others would lead to an absolute misinterpretations. Another common confusion regarding interpretation of regression parameters was due to neglecting a vital fact about the analysis: *"Regression or correlation does not imply causation"*. Therefore, in this manuscript, it was avoided to interpret the results of the models with verbal expression such "... an increase in X would *cause* also an increase in Y".

In addition, the models were collinear as mentioned previously. Consequently, one of the estimated parameters of two highly correlated variables was expected to be negatively signed. An explicit example of this situation could be observed for excavation volume (m310) and internal walls area (m340) models, which both retained gross external floor area (GEFA) and gross building volume (GBV) at the same time in the model. Consequently, parameter of gross building volume was estimated to be negative (Table 5.2). Initially, this offenses common sense. However, the parameter, was also called *partial regression coefficient,* of gross building volume was negatively signed because *the parameter* represented the contribution of gross building volume *after it was adjusted* for other predictor variables.

Table 5.2: Estimated parameters and closeness of fit for final linear regression models of reference unit quantities

Var.	Final model responses					
	$m310^{0.21}$	$m320^{0.85}$	$m330^{0.60}$	$m340^{0.50}$	$m350^{0.89}$	$m360^{0.72}$
GEFA	0.00027	0.0096	0.01131	0.00829	0.3457	0.0056
GBV	−0.00004			−0.00067		
NoS			5.52124			
NoS_{ag}	0.08605			3.10747		
NoS_{bg}				2.71640	59.5652	
PA			0.00016	0.00013		
DPA		0.1995	−0.00183		−0.2739	0.0396
ASH		7.5821	4.13878			5.9416
AFS	0.00080	0.0609	0.00480	0.00408	−0.0844	0.0508
LI	1.28720					
FAC	β_c^{m310}	β_c^{m320}	β_c^{m330}	β_c^{m340}	β_c^{m350}	β_c^{m360}
GC						
SA						
TOP	0.24540		1.76750		−15.8063	
Constant	*1.55223*	*10.1360*	*1.32153*	*7.71147*	*64.0677*	*19.0358*
R^2	69.67%	98.74%	93.56%	91.83%	98.31%	95.73%

Facility type served as an adjustment for the regression intercept. This is to say, its interaction with other hypothesized variables was *not* assumed in the linear models. Selecting *'residential'* projects as the control group, model *constant* was the value of model intercept for residential projects. In order to adjust model intercept according to other types of facilities, one would add estimated parameters for other types of buildings (indicator variables for different levels) to the control group constant. These parameters were represented by $\beta_c^{m310}, \beta_c^{m320}, \beta_c^{m330}, \beta_c^{m340}, \beta_c^{m350}, \beta_c^{m360}$ in Table 5.2. Estimated parameters of different facility types also provided a fair ground to compare in between average expected values of reference unit quantities for different facility types. For instance, associated model (m330) suggested that expected average value of external walls area for office developments was significantly greater than residential ones for the population of interest, German building industry, after adjusted with the other significant factors.

Variation in excavation volume (m310) could best be explained by a linear combination of gross external floor area (GEFA), gross building volume (GBV), number of storeys above ground (NoS_{ag}), average floor size (AFS), location index (LI), facility type (FAC) and topographical conditions (TOP). The model was able to cover approximately 70% of variance in excavation volume. This corresponded to a moderate fit. Although it showed high relevance in practice, the variable associated with number of storeys below ground made an insignificant contribution to the model after adjusted with predictors retained in the final model (Table 5.2). Therefore, it was dropped from the final model. It might be that the variance which was assumed to be explained by number of storeys below ground, could also be explained by facility type and topographical conditions. This might be the motivation why the variable became redundant for the model. Coefficients of indicator variables were estimated as $\beta_c^{m310} = (0.12396, 0.29146, 0.55735, 0.67196)$ for education, industrial, office and other facility types, respectively. The result suggested that average excavation volume of residential buildings were smallest followed by educational, industrial, office and others after adjusted with other significant factors.

According to the results of ordinary least square parameter estimation, one could see in

Table 5.2 that gross external floor area (GEFA), developed plot area (DPA), average storey height (ASH), average floor size (AFS), and facility type (FAC) were the significant drivers of foundation area model (m320). The final model covered approximately 98% of the variance in the foundation area. This was *more than ideal*. Analysis revealed gross building volume (GBV), number of storeys above ground (NoS_{ag}), and ground conditions (GC) did not explain significant portion of variance in the model response. The amount of variance by gross building volume might be covered by gross external floor area and their combination became redundant for the model. Hence, gross external floor area covered more variance and showed more practical relevance it was retained in the model. Number of storeys above ground was significantly related to the average storey height and this might cause an insignificant t score for the parameter estimation. Ground condition initially offered practical relevance. But when one investigated the data in detail, two underlying reasons might be present why this variable became redundant: First, it was more relevant to the cost of foundation than the amount of it. Ground condition was a factor that may affect the unit price of foundation area but not the amount. Second, among observations the value showed similarity and therefore as a variable it did not help us to distinguish between observations. Coefficients of indicator variables of facility type were estimated as $\beta_c^{m320} = (15.6272, 19.5235, 7.5766, 44.7236)$ for education, industrial, office and other facility types, respectively. The result suggested expected average foundation area of residential developments were the smallest when compared to other facility types. Residentials were followed by offices, education, industrial and other facility types, after adjusted with the significant factors of *m320* model.

Predicting external walls area accurately showed greater practical relevance because it was the reference unit quantity of a cost group which had the highest amount in many cases among other cost groups of structure. Developed prediction model for external walls area (m330) suggested gross external floor area (GEFA), number of storeys (NoS), plot area (PA), developed plot area (DPA), average storey height (ASH), average floor size (AFS), facility type (FAC), and topographical conditions (TOP) provided the best linear combination of candidate predictors which explained the highest portion of variance in the response (Table 5.2). Gross building volume and ground conditions had insignificant contribution to explained variance in the final model. Ground conditions did not show a practical relevance, On the other hand, explained variance by the gross building volume was probably covered by gross external floor area. Parameter of developed plot area was estimated to be negative due to collinearity as discussed previously. The final model covered approximately 94% of the variance in external wall area, which might offer an *ideal* prediction accuracy of unseen data. Coefficients of indicator variables of facility type were estimated as $\beta_c^{m330} = (9.64954, 3.56462, 6.09710, 7.11390)$ for education, industrial, office and other facility types, respectively. The result suggested expected average external walls area of residential developments were the smallest when compared to other facility types. Residentials were followed by industrial, offices, others and educational, after adjusted with the significant factors of *m330* model.

Developed model of internal walls area (m340) offered gross external floor area (GEFA), gross building volume (GBV), number of storeys above and below ground (NoS_{ag}, NoS_{bg}), plot area (PA), average floor size (AFS), and facility type (FAC) was the best subset of variables, which explained the highest variance in response, out of all assumed predictors available at HOAI basic evaluation stage. Developed plot area was redundant, given the fact it was linearly dependent to average floor size, and plot area significantly. Thereof, its contribution to the model was insignificant and can be neglected. Average storey height, on the other hand, showed *singularity* with gross external floor area and gross building volume. Presumably, this variable attempted to cover identical part of variance in the internal walls area and became redundant as well. Lastly, site access, an indicator of a project location to an extent, was dropped from the model. The relationship between facility type and site access might have caused this as well. The final model was able to cover around 92% of variance in internal walls area (Table 5.2). Coefficients of indicator variables of facility type were estimated as $\beta_c^{m340} = (4.28282, -1.85357, 0.73209, 2.74707)$ for education, industrial, office and other facility types, respectively. Unlike others, the result suggested expected average internal walls area of industrial developments were the smallest when compared to other facility types. Industrials were followed by residentials, offices, others and educationals, after adjusted with the significant factors of *m340* model.

Floors and ceilings area model (m350) revealed linear combination of gross external floor area (GEFA), number of storeys below ground (NoS_{bg}), developed plot area (DPA), average floor size (AFS), facility type (FAC) and topographical conditions (TOP) provided the highest amount of explained variance in the model response. Initial candidate predictors of gross building volume (GBV), plot area (PA), average storey height (ASH) and ground conditions (GC) did not contribute the model significantly and therefore excluded from the final model. Presumably, these variables accounted for the same portion of explained variance which was already covered by final model predictors, like in other models. Likewise, parameters of developed plot area and average floor size were estimated negatively due to collinearity. The final model accounted for 98% of variance in the model response (Table 5.2). Coefficients of indicator variables of facility type were estimated as $\beta_c^{m350} = (23.0378, -12.1946, 53.1425, -23.1718)$ for education, industrial, office and other facility types, respectively. A different pattern was observed. The result suggested expected average floors and ceilings area of other type of developments, such as cultural buildings, are the smallest when compared to other facility types. Industrial buildings were also smaller than residential buildings in terms of expected average floors and ceilings area. Residentials were ranked as number three followed by education and office developments, after adjusted with the significant factors of *m350* model.

Roof area was the last reference unit quantity model (m360), which contained gross external floor area (GEFA), developed plot area (DPA), average storey height (ASH), average floor size (AFS), and facility type (FAC) as the predictors at 5% significance level. Gross biding volume (GBV) and number of storeys above ground (NoS_{ag}) were the variables initially included in the full model. Analysis revealed their contribution to the model was not significant and therefore dropped form the model. The final model accounted for approximately 96% of variance in the roof area. Coefficients of indicator variables of facility type were estimated as $\beta_c^{m360} = (10.5191, 1.0903, 8.4835, 24.4394)$ for education, industrial, office and other facility types, respectively. The result suggested that average expected roof area of residential were the smallest when compared to the other types of facilities. This was followed by industrials, offices, educationals and others, respectively, after adjusted with the significant drivers of *m360* model.

One could see in Table 5.2 that gross external floor area (GEFA), average floor size (AFS) and facility type (FAC) were three determinants that were retained in all final reference unit quantity models. Recall that these variables were also identified as candidate predictors for all initial models by industry practitioners and stepwise regression procedure. This indicated that the results presented practical relevance. Previously in Chapter 1, it was indicated that our main hypotheses rely on a pilot investigation of the data set and detection of high correlation in between reference unit quantities and gross external floor area. Thereof, this was expected. Initially, average floor size was assumed to have a strong pattern with horizontal building elements such as foundation and roof area. Yet, its contribution was significant for all models after adjusted with other significant drivers. Lastly, facility type, which was assumed as one of the most important and available information in the data set at the early stages of a development, explained a substantial amount of variance in the response variables. Moreover, it added a valuable contribution to the interpretation of analysis by providing a ground for comparative results of mean expected average values in between different facility types. Gross building volume (GBV), which was also listed as a candidate predictor, was dropped from the models of foundation area (m320), external walls area (m330), floors and ceilings area (m350) and roof area (m360). It was argued that gross building volume aimed to account for the part of variation in the response variables which was also covered by other predictor variables, in this case especially gross external floor area.

Table 5.3: Prediction performance of linear regression models of reference unit quantities

Performance measure	Model					
	m310	m320	m330	m340	m350	m360
RMSE	4005.94	305.04	554.47	1283.37	441.04	347.14
MAPE	52.22%	15.34%	23.65%	38.82%	27.38%	22.10%
$n*$	6	0	1	3	3	0

Note: "$n*$" stands for number of outliers in the test sample.

A good fit of the model to the train sample was, of course, vital for more accurate prediction; yet might not be enough. One would test predictive ability via using developed models to predict the values of interests for unseen data, *the test sample*. Table 5.3 showed average values for prediction performance indicators of reference unit quantity models on the *test sample* that contains 57 observations (Chapter 4). As expected, due to its relatively low fit to the train data, excavation volume (m310) model reached to the highest error, followed by internal walls area (m340), both provide average prediction errors greater than 30% for unseen data. In addition, higher number of projects were detected as outliers in the test sample when these models were considered. The rest of the models provided errors that were lower than 30% as well as observed decrease in number of outliers for the test sample. It was concluded the prediction performances of the models were *ideal* for the reason that the primary aim of the study was not to provide prediction of reference unit quantities but to employ their fit values to enhance the predictions of cost of structure and construction duration.

Another method of analysis, artificial neural networks, was implemented to enable comparative results with linear regression models. Repeatedly, the aim was not set to develop an artificial neural network software, rather to test their pattern recognition and prediction capability for the problem of interest, in this case project determinant prediction. Thereof, a commercial software was utilized. Computation steps such as variable transformation, calculating arc weights, and so on, were not detailed in the manuscript for the aim of clarity. Those who are interested in details of these calculations is referred to the study by Hegazy & Ayed (1998) in which they present a spreadsheet implementation of neural network step by step to estimate cost of highway projects in USA.

As indicated in Chapter 3, the input nodes of ANN models were identical to those of linear regression models which can be tracked in Table 5.2. Indeed, only significant predictors of reference unit quantities were considered. Adjustment of network architecture and determination of optimum number of training iterations were decided heuristically. Starting from 30,000 and increasing 10,000 each time until 100,000 number of iterations were used to determine optimum number of iterations. It was observed that 50,000 training iterations were generally obtained as threshold values. After this point the network model tended to over-train which limited their prediction capability significantly. Repeatedly, this was a heuristic process and we strictly did not argue that it would result in the same way with another data set having different number of observations and predictors.

Secondly, it was aimed to determine the best type of neural network and, if necessary then, the architecture for the reference unit quantity models. For each model, *general regression neural* (GRN) nets and *multi layer feedforward* (MLF) nets having 2 to $2n+1$, where n stands for number of input variables, number of nodes in its hidden layer were trained. The model that minimized root mean squared error of the test sample, contained 57 objects, was selected as the most adequate one to predict values of interest in the unseen data.

Table 5.4: Architecture and prediction performance of ANN models for reference unit quantities

ANN architecture and performance parameters	Model					
	m310	m320	m330	m340	m350	m360
GRN Net				■		
MLF Net	■	■	■		■	■
No. nodes in hidden layer	3	2	4	-	3	4
RMSE of best configuration	2279.17	300.13	422.30	854.96	329.56	365.19
MAPE of best configuration	55.13%	11.44%	17.41%	38.19%	18.13%	18.40%
n^*	7	0	1	3	5	1

Note: "n^*" is the number of outliers and/or zeros in the test group

The results were presented in Table 5.4. Accordingly, one could see that for all models, except the one associated with internal walls area (m340), MLF Nets provided superior prediction accuracy. Our findings suggested that internal wall area could best be predicted by a GRN Net, which did not require any setting for its architecture. Detecting 3 objects as outliers internal walls area model reached to 38% mean error. Result of prediction capability showed no difference when compared to linear regression models for internal walls area when MAPE was considered as the measure (Table 5.3). Yet, it was noted that root mean square error (RMSE) of ANN model was significantly lower than linear regression model (LRM).

Best excavation volume model (m310) was selected as the MLF Net having 3 nodes in its hidden layer. 7 buildings in the test sample were identified as outliers. MAPE was computed as 55% over remaining 50 observation in the test sample. This was a slight increase compared to LRM. However, ANN model provided significantly smaller value of RMSE than LRM, if RMSE was considered as the measure of prediction accuracy (Table 5.3 and 5.4).

The neural net model which minimized the error of test sample for foundation area (m320) was determined also as the MLF Net but this time with 2 nodes in its hidden layer. The ANN model successfully covered the variance in the test sample as no outlier was detected like in LRM. When ANN and LRM model was employed, the test sample reached 11% and 15% of error, respectively. On the other hand, RMSE of two models did not show disparity (Table 5.3 and 5.4).

External walls area (m330) of the test sample was predicted with minimum error by the MLF Net with 4 neurons in its hidden layer. Like LRM, ANN model detected one outlier in the test sample. MAPE was calculated nearly 6% less than LRM over 56 remaining objects in the test sample. Like MAPE, RMSE was also computed approximately 20% less than LRM (Table 5.3 and 5.4).

The MLF Net with 3 nodes in its hidden layer was the one which minimized RMSE of test sample for floors and ceilings area (m350). The model reached to MAPE of 18% over test sample. This was significantly smaller than its counterpart, LRM. However, ANN model detected 5 observations as outliers in test sample while this number was 3 for LRM. RMSE values, on the other hand, also favor ANN model over LRM. Yet, relatively higher number of outliers in ANN model might be an indicator of lower capability of pattern recognition for floors and ceilings area (Table 5.3 and 5.4).

Lastly, prediction error of roof area over the test sample was minimized by the MLF Net with 4 neurons in its hidden layer. Detecting only one observation as an outlier, ANN model reached MAPE of 18%. This indicated 4% improvement when compared to LRM. However, when RMSE was considered as the measure of accuracy, one could see in Table 5.3 and 5.4 that LRM result was slightly better, having no outlier in the test sample.

In conclusion the results revealed, LRM and ANN models did not show significant differences of prediction accuracy for excavation volume (m310) and internal walls area (m350), when considering MAPE as the measure of accuracy. However, the findings suggested, ANN models provided superior prediction accuracy when compared to LRM for the rest of reference unit quantities, namely foundation area (m320), external walls area (m330), floors and ceilings area (m350) and roof area (m360). Considering RMSE as the performance indicator, ANN models were superior to their counterpart, LRM. Only RMSE of ANN model of roof area (m360) was found slightly greater than LRM model.

Implementation of the resulted models on a randomly selected case building was demonstrated exhaustively in Chapter 7. One could find the method of constructing linear regression equations out of Table 5.2, and confidence interval estimations in relevant sections of Chapter 7.

5.2 Cost of Structure

Construction cost estimation was performed in accordance with the hierarchical structure of relevant German norm, DIN 276-1 and hypotheses to be tested. Thereof, from now on, construction cost would be referred to as cost of structure as suggested by DIN 276-1 (Figure 1.2). Cost estimation was performed at three levels and *bottom to top* approach was adopted to present the results.

According to constructed frameworks (Figure 1.4), which form the primary hypothesis of interest regarding cost estimation, second level cost groups of structure was modeled, first. This, Framework #1, corresponded to feedback system inputs with the predicted values of reference unit quantities (Section 5.1), which in turn might increase the portion of explained variance in the second level cost groups, such as cost of external walls (k330). Following, Framework #2, which suggested feeding back system inputs with reference unit quantities, as framework #1, to estimate first level cost groups of structure (i.e. k300), was employed. Thirdly, Framework #3 that offered using only the variables available to an architect at HOAI basic evaluation stage to estimate first level cost of structure (i.e. k400) was followed. Lastly, framework #4, which utilized limited predictors as framework #3 to estimate total cost of structure (k300+k400).

5.2.1 Second level cost groups of structure

Prior to detailed explanation of developed models, restrictions applied to data set would be underlined. The initial training and test sample, which were used for modeling and measuring prediction accuracy aims, consisted of 600 and 57 objects, respectively. Through analysis of reference unit quantities, inevitably, some observations were detected as outliers and excluded from both train and test sample. Therefore, the fit (predicted) values of these observations became missing values when the system inputs were fed back with the predicted values of reference unit quantities. These missing values could not be employed in further analysis. Moreover, six of each reference unit quantity models detected different objects as outliers. In conclusion, when filtering applied to the data set, which contains predictors of not only initial information but also fit values of reference unit quantities, 475 and 49 observations remained in train and test sample, respectively. Recall that, it was assumed the architect was capable of assessing the second level cost groups which do not have any value (zero).

Methodological principles, as outlined in Section 5.1, applied through reference unit quantity modeling for both linear regression and artificial neural network were also implemented to cost estimation. Therefore, aiming the clarity of the manuscript, we refer readers to associated section for details.

Table 5.5 presented the predictor variables that were included in the initial models. Stepwise regression method was implemented to detect the variables that would retain in the initial models of second level cost groups of structure. The stepwise regression method is essentially a forward selection procedure but with the added proviso that each stage the possibility of deleting a variable, as in backward elimination is considered. 10% significance level was selected.

Considering models associated with the second level cost groups of structure, construction works - such as cost of roof - two categorical variables, namely facility and client type (FAC, CT), appeared in all initial models. These variables were followed by gross building volume (GBV) and average floor size (AFS) which were hypothesized to be significant drivers of 7 and 6 cost groups, respectively. Special attention was paid to the reference unit quantities and their contribution. Parallel to main hypothesis, initial analysis demonstrated that reference unit quantities of second level cost groups were relevant at least at 10% significance level. It shall also be underlined that, when reference unit quantity of a cost group was retained, gross external floor area (GEFA) was become insignificant for the models with two exceptions: external and internal walls area models (m330 and m340). This indicated that reference unit quantities accounts for greater portion of variance in the second level cost groups of construction works when compared to gross external floor area (Table 5.5).

The figure changes dramatically as gross external floor area (GEFA) appeared in 8 of 9 models initially when second level cost groups of structure, services - such as cost of sewerage, water and gas systems - were observed attentively. Parallel to findings of stepwise selection procedure, DIN 276-1 indicated reference unit quantities of these cost groups as gross external floor area. Similar to construction works models, facility type (FAC) was a candidate predictor for all models of services. It should also be noted that predicted values of external walls area ($m330_{fit}$) were initially appeared in 7 of second level cost groups of service models (Table 5.5). Similar fitted values of excavation volume ($m310_{fit}$), internal walls area ($m340_{fit}$), floors and ceilings area ($m350_{fit}$) and roof area ($m360_{fit}$) were appeared in 5, 6, 3, and 1 time(s) in initial models for second level cost of services, respectively.

Collective results of ordinary least square parameter estimation for second level cost groups, construction works were presented in Table 5.6. One can see, all responses were transformed to fulfill model assumptions of linear regression models.

The results suggested the variation in cost of excavation (k310) could best be explained by the linear combination of fitted values of excavation volume ($m310_{fit}$) and foundation area ($m320_{fit}$), gross building volume (GBV), number of storeys below ground (NoS_{bg}), developed plot area (DPA), facility type (FAC), site access (SA), and market conditions (MC). Due to the high correlation in between fit value of foundation area and developed plot area, the parameter of foundation area was reported as negatively signed. Site access (SA) revealed the information regarding the surroundings of the development to be built. Thereof, this variable showed practical relevance. It can be argued, the cost of excavation tended to increase significantly due necessity of extra support systems, when the excavation had to be conducted next to

Table 5.5: Candidate predictors of second level cost groups of structure models

Var.	Initial model in which the factor is included																
	k310	k320	k330	k340	k350	k360	k370	k390	k410	k420	k430	k440	k450	k460	k470	k480	k490
GEFA			■	■			■		■	■		■	■	■	■	■	■
GBV	■	■	■	■	■		■	■			■	■	■		■		
NoS			■		■	■					■		■				
NoS_{ag}			■	■			■	■		■		■			■		
NoS_{bg}	■					■	■	■	■								
PA		■			■	■						■	■			■	
DPA	■		■		■	■				■			■		■		
ASH	■		■	■	■			■	■		■	■	■		■		
AFS		■	■	■	■	■		■	■	■	■	■	■	■	■		
LI			■		■					■		■	■			■	
FAC	■	■	■	■	■	■	■	■	■	■	■	■	■	■	■	■	■
GC		■			■										■		
SA	■		■		■	■											
TOP		■	■	■	■			■			■	■	■			■	
MC	■	■		■		■						■				■	
CT	■	■	■	■	■	■	■	■	■	■	■	■	■	■	■		
$m310_{fit}$	■		■	■	■	■			■		■	■	■	■			
$m320_{fit}$	■	■	■		■	■											
$m330_{fit}$			■	■	■			■	■		■	■	■	■	■	■	
$m340_{fit}$	■		■	■	■			■	■		■	■	■	■	■		
$m350_{fit}$					■		■		■			■	■				
$m360_{fit}$			■	■	■	■									■		

an existing building. Developed plot area and foundation area were two indicators of section area for excavation volume to an extent and also shows practical relevance. Gross building volume in combination with number of storeys below ground indicated the amount of volume that is under the ground level also presents significance relationship with cost of excavation. In addition it can be argued, this combination influenced average storey height (ASH) and internal walls area (m340$_{fit}$), which were initially included in the model, became redundant for the final model. The final model accounted for 58% of variance in the cost of excavation, indicating a moderate fit that was *less than ideal* to predict unseen data (Table 5.6). Coefficients of indicator variables of facility type were estimated as $\beta_c^{k310} = (0.16974, 0.18082, 0.23177, 0.39920)$ for education, industrial, office and other facility types, respectively. The result suggested that average expected excavation cost of residentials were the smallest when compared to the other types of facilities. Residentials were followed by education, industrial, office and other facility types, respectively, after adjusted with the significant drivers of *k310* model.

Second column of the Table 5.6 summarized the linear model associated with cost of foundation (k320). According to results, fit of foundation area (m320$_{fit}$), gross building volume (GBV), plot area (PA), facility type (FAC), and market conditions (MC) were the significant predictors of cost of foundation. All retained variables contributed at least 5% significance level to the model. Yet, it would be underlined, in consistent with analysis of variance, foundation area in combination with facility type took the lead role in this model and explained substantial part of variance when compared to other retained variables. Ground conditions (GC) was hypothesized to be included in the model and showed high level of practical relevance. Yet, when one investigated the data set carefully many objects were classified as BK 3 to 5 which corresponded to moderate. This might be the reason why GC was not able to cover a significant portion of variance in the response variable and dropped from the final model. By other words it could be argued, having so many objects with the same value, the variable lost its power to distinguish in between projects having different cost of foundation. The results also denied topography class (TOP), and client type (CT) as significant variables of the final model. Amount of variance that could be covered by the final model was approximately 88%. This indicated a good fit and can be classified as *ideal* for prediction aims. Coefficients of indicator variables of facility type were estimated as $\beta_c^{k320} = (19.8560, 12.1103, 16.8736, 23.9999)$ for education, industrial, office and other facility types, respectively. The result suggested that average expected cost of foundation for residentials are the smallest when compared to the other types of facilities. Residentials were followed by industrial, offices, education and other facility types, respectively, after adjusted with the significant drivers of *k320* model.

Analysis revealed the linear model, consisted of fitted values of external walls, internal walls, and roof area (m330$_{fit}$, m340$_{fit}$, m360$_{fit}$), number of storeys and number of storeys above ground (NoS, NoS$_{ag}$), developed plot area (DPA), average floor size (AFS), facility and client type (FAC, CT), accounts for the greatest portion of variance in cost of external walls (k330). It was argued that due to high correlation observed, fitted values of roof and internal walls area parameters were estimated as negative. A similar figure observed between number of storeys above ground and number of storeys. Although these variables caused substantial amount of collinearity problem for the model, their contribution to the model and therefore to the prediction was significant, which was evident by the statistics. Therefore, these variables were retained in the final model. Recall, interpretation based on parameters of these variables would be conducted, considering the other significant variables in the model. On the other hand, the information by gross external floor area (GEFA), and gross building volume (GBV) became redundant after including three building elements quantities in the model. It could be argued, average storey height (ASH) tended to explain the same part of variance as variables associated with number of storeys and therefore its contribution was insignificant, too. Lastly, analysis suggests location index (LI), site access (SA), and topographical conditions (TOP) either did not explain substantial amount of variance in the response or the information was already covered by

Table 5.6: Estimated parameters and closeness of fit for final linear regression models of second level cost groups of structure, construction works

Var.	Final model responses							
	$k310^{0.11}$	$k320^{0.37}$	$k330^{0.07}$	$k340^{0.33}$	$k350^{0.39}$	$k360^{0.11}$	$k370^{0.16}$	$k390^{0.15}$
GEFA				0.0183				
GBV	0.00003	0.0013		−0.0023			0.00005	
NoS			0.03667		14.7170	0.05278		
NoS_{ag}			−0.01625				−0.20091	0.19533
NoS_{bg}	0.26379						0.32101	0.18358
PA		0.0002			0.0006	2.85E-0.06		
DPA	0.00125		0.00028			0.00314		
ASH								
AFS			0.00035			0.00030		0.00038
LI								
FAC	β_c^{k310}	β_c^{k320}	β_c^{k330}	β_c^{k340}	β_c^{k350}	β_c^{k360}		β_c^{k390}
GC								
SA	0.12981				2.7861			
TOP								
MC	0.07088	2.3154		2.1532		0.06552		
CT			β_p^{k330}	β_p^{k340}	β_p^{k350}	β_p^{k360}	β_p^{k370}	β_p^{k390}
$m310_{fit}$	0.00003				0.0057	0.00003		
$m320_{fit}$	−0.00144	0.0163				−0.00460		
$m330_{fit}$			0.00023	0.0244	−0.0263			0.00081
$m340_{fit}$			−0.00008	−0.0168	−0.0356			−0.00029
$m350_{fit}$					0.0672		0.00023	
$m360_{fit}$			−0.00051		0.0278	0.00133		
Constant	*2.22456*	*39.8236*	*2.26970*	*23.4094*	*19.0973*	*3.10086*	*4.62793*	*3.27456*
R^2	57.95%	87.83%	85.87%	89.88%	91.56%	85.69%	48.14%	68.80%

another significant factor(s). The final model accounted for 86% of variance in cost of external walls, which can be regarded as *ideal* for prediction aims (Table 5.6). Coefficients of indicator variables of facility type were estimated as $\beta_c^{k330} = (0.07083, -0.02155, 0.10331, 0.12050)$ for education, industrial, office and other facility types, respectively. The result suggested that average expected cost of external walls for industrial projects were the smallest when compared to the other types of facilities. Industrials were followed by residentials, educational, offices and other facility types, respectively, after adjusted with the significant drivers of *k330* model. Lastly, coefficient of public buildings was estimated as $\beta_p^{k330} = (0.03437)$, which indicated a higher average expected value when compared to private developments.

The result of regression equation for cost of internal walls (k340) suggested that the greatest portion of variance in the response could best be covered by the linear combination of fitted values of internal and external walls ($m340_{fit}$, $m330_{fit}$), gross external floor area (GEFA), gross building volume (GBV), facility and client type (FAC, CT), and market conditions (MC). Like in other models which contained internal and external walls area at the same time, estimated parameter of internal walls area was negative due collinearity. The same pattern observed in between variables gross external floor area and gross building volume. Estimated parameter of gross building volume was also negative. It could be argued, variation that could be explained by number of storeys above ground (NoS_{ag}), average storey height (ASH) and average floor size (AFS) could also be covered by the variables that were retained in the final model, such as gross external floor area and external walls area. In addition, topography (TOP), fit of excavation volume ($m310_{fit}$) and roof area ($m360_{fit}$) did not show any practical relevance and therefore dropped from the model. The final model accounted for 90% of variance in cost of internal walls which could be regarded as *ideal* for prediction aims (Table 5.6). Coefficients of indicator variables of facility type were estimated as $\beta_c^{k340} = (4.8774, -6.4124, 4.4668, 4.8386)$ for education, industrial, office and other facility types, respectively. The result suggested that average expected cost of internal walls for industrial projects were the smallest when compared to the other types of facilities. Industrials were followed by residential, office, other and education facility types, respectively, after adjusted with the significant drivers of *k340* model. Lastly, coefficient of public buildings was estimated as $\beta_p^{k340} = (3.0355)$ which indicates significantly higher average expected value when compared to private developments.

The determinants of the linear model that explained the highest portion of variance in cost of floors and ceilings (k350) were *fitted* values of floors and ceilings ($m350_{fit}$), roof ($m360_{fit}$), internal ($m340_{fit}$) and external walls area ($m330_{fit}$), excavation volume ($m310_{fit}$), number of storeys (NoS), plot area (PA), site access (SA), facility (FAC) and client type (CT). Linear combination of reference unit quantities explained the substantial variance in cost of floor and ceiling (Table 5.6). This is why macro project determinants that were initially hypothesized to be relevant, namely gross building volume (GBV), average storey height (ASH), average floor size (AFS) became redundant and therefore were not retained in the final model. Although location index (LI) showed practical relevance, statistical evidence suggested its contribution was insignificant at 5% level. Ground conditions (GC) and topographical properties (TOP) dropped from the final model due to their insignificant regression parameters. Also, these variables (GC and TOP) did not provide practical relevance to explain any variance in the response. Lastly, fitted value of foundation area ($m320_{fit}$) presumably tend to explain the same part of variance as fitted value of floors and ceilings area. Thereof, its contribution to regression was limited and the variable became insignificant. The final model accounts for 92% of variance in cost of floors and ceilings, which can be regarded as *more than ideal* for prediction aims. Coefficients of indicator variables of facility type were estimated as $\beta_c^{k350} = (19.2971, 0.2806, 12.0236, 8.2402)$ for education, industrial, office and other facility types, respectively. The result suggested that average expected cost of floors and ceilings for residential projects were the smallest when compared to the other types of facilities. Residentials were followed by industrial, cultural, office and education facility types, respectively, after adjusted with the significant drivers of *k350* model. Lastly, coefficient of public buildings was estimated as $\beta_p^{k350} = (9.0209)$ which indicated substantial higher value of expected average cost of floors and ceilings when compared to private developments.

The last model that directly associated with its reference unit quantity was cost of roof (k360). Result of regression analysis suggested linear combination of fitted values of roof area ($m360_{fit}$), foundation area ($m320_{fit}$), and excavation volume ($m310_{fit}$), along with number of storeys (NoS), plot area (PA), developed plot area (DPA), average floor size (AFS), market conditions (MC), facility (FAC) and client type (CT) explained the greatest possible variance in cost of roof (Table 5.6). It is noticeable, many significant predictors retained in the final model were associated to the horizontal cross section of the building such as developed plot area, foundation area, and average floor size. The only vertical measurement of a building that was found to be significant was number of storeys. Two variables were included in the initial model, namely number of storeys below ground (NoS_{bg}) and site access (SA). They were excluded from the final model due practical irrelevance and statistical insignificance. The final model explained 86% of variance in cost of roof, which could be identified as *ideal* for prediction aim. Coefficients of indicator variables of facility type were estimated as $\beta_c^{k360} = (0.18484, 0.29704, 0.17614, 0.48828)$ for education, industrial, office and other facility types, respectively. The result suggested that average expected cost of roof for residential projects were the smallest when compared to the other types of facilities. Residentials were followed by office, education, industrial and other facility types, respectively, after adjusted with the significant drivers of *k360* model. Lastly, coefficient of public buildings was estimated as $\beta_p^{k360} = (0.09609)$ which indicated slightly higher value of expected average cost of roof when compared to private developments.

Analysis indicated only 47% of the variance could be explained in cost of fitments permanently fixed to the building fabric (k370) by the linear regression model (Table 5.6). Fitted value of floors and ceilings area ($m350_{fit}$), gross building volume (GBV), number of storeys below ground (NoS_{bg}), number of storeys above ground (NoS_{ag}), and type of client (CT) were the significant predictors. Results suggested gross external floor area (GEFA) and facility type (FT) did not provide a significant contribution to explain variance in the model response and therefore excluded from the final model. Coefficient of public buildings was estimated as $\beta_p^{k370} = (0.57516)$ which indicated higher value of expected average cost of structural fitments when compared to private developments.

The last model of second level cost of structure, construction works, was cost of other construction related activities such as scaffolding (k390). Results by regression analysis suggested number of storeys above (NoS_{ag}) and below ground (NoS_{bg}), average floor size (AFS), facility (FAC) and client type (CT), fitted values of external ($m330_{fit}$) and internal walls ($m340_{fit}$) were the best subset of predictors that explained approximately 69% of variance in cost of other construction related activities (Table 5.6). It could be argued, the combination of external and internal walls area aimed to account for the same portion of variance as gross building volume; thus the variable became redundant and excluded from the final model. Similarly, variance that could be explained by average storey height (ASH), presumably could also be covered by number of storey related variables. Thus, ASH were also dropped from the final model. Coefficients of indicator variables of facility type were estimated as $\beta_c^{k390} = (0.40956, 0.06975, 0.60413, 0.40952)$ for education, industrial, office and other facility types, respectively. The result suggested that average expected cost of other construction related works for residential projects were the smallest when compared to the other types of facilities. Residentials were followed by industrial, others, educational and office facility types, respectively, after adjusted with the significant drivers of *k390* model. Lastly, coefficient of public buildings was estimated as $\beta_p^{k390} = (0.19629)$ which indicated slightly higher value of expected average cost of other construction related activities when compared to private developments.

Results of ordinary least square estimation of nine second level cost groups which aggregated total cost of structure, services (k400) were presented in Table 5.7. Identical to the models associated with second level cost groups of structure, construction works, all response variables were subject to variable transformation to fulfill assumptions of linear regression modeling. Investigation started from the first one, cost of sewerage, water and gas systems (k410). The results revealed the final model that consisted of gross external floor area (GEFA), number of storey below ground (NoS_{bg}), facility (FAC) and client type (CT), fitted values of excavation volume ($m310_{fit}$) and external walls area ($m330_{fit}$) could account for 80% of variance in cost of sewerage, water and gas systems (Table 5.7). Due to collinearity between gross external floor area and fitted values of reference unit quantities, parameters of $m310_{fit}$ and $m330_{fit}$ were estimated as negative. Parameters of average floor size (AFS) and average storey height (ASH) became insignificant after adjusted with variables retained in the final model. It could be argued that the portion of variance that could be covered in the response by AFS and ASH, could also be covered by gross external floor area. It was evident by the statistical analysis that fitted values of internal walls ($m340_{fit}$) and ceilings area ($m350_{fit}$) also became redundant for the final model. The reason might be existing correlation between these variables and gross external floor area. On top, parallel to practice, it was observed that the biggest portion of variance in k410 model was explained by gross external floor area (Table 7.5). This result statistically justified the motivation stood for pointing gross external floor area as reference unit quantity of cost of sewerage, water and gas systems. Coefficients of indicator variables of facility type were estimated as $\beta_c^{k410} = (0.14175, -0.10225, 0.06100, 0.53870)$ for education, industrial, office and other facility types, respectively. The result suggested that average expected cost of sewerage, water and gas systems for industrial projects were the smallest when compared to the other types of facilities. Industrials were followed by residential, office, educational and other facility

Table 5.7: Estimated parameters and closeness of fit for final linear regression models of second level cost groups of structure, services

Var.	Final model responses								
	$k410^{0.16}$	$k420^{0.22}$	$k430^{0.10}$	$k440^{0.12}$	$k450^{0.05}$	$k460^{0.36}$	$ln(k470)$	$k480^{0.5}$	$ln(k490)$
GEFA	0.00081	0.00072		0.00068	0.00016		0.00095	0.046	0.00037
GBV			−0.00003	−0.00009					
NoS					0.01240				
NoS_{ag}		0.16890					−0.25577		
NoS_{bg}	0.25069								
PA				0.00001				0.011	
DPA		0.00060			−0.00010				
ASH				0.08297					
AFS			0.00031	−0.00019		−0.0091			
LI		2.85717						−785.332	
FAC	β_c^{k410}	β_c^{k420}	β_c^{k430}	β_c^{k440}	β_c^{k450}	β_c^{k460}			
GC									
SA									
TOP				−0.06414					
MC				0.05297				100.092	
CT	β_p^{k410}		β_p^{k430}	β_p^{k440}	β_p^{k450}	β_p^{k460}			
$m310_{fit}$	−0.00015								
$m320_{fit}$									
$m330_{fit}$	−0.00054		0.00076	0.00088		0.0211			
$m340_{fit}$			−0.00025	−0.00056		−0.0066	−0.00058		
$m350_{fit}$				−0.00020	−0.00015				
$m360_{fit}$									
Constant	*5.22266*	*6.23933*	*2.03938*	*2.57729*	*1.41826*	*40.5248*	*8.70902*	*664.657*	*7.22284*
R^2	79.75%	77.23%	53.61%	82.91%	66.72%	63.72%	32.55%	88.43%	16.05%

types, respectively, after adjusted with the significant drivers of *k410* model. Lastly, coefficient of public buildings was estimated as $\beta_p^{k410} = (0.25923)$ which indicated higher value of expected average cost of sewerage water and gas systems when compared to private developments.

The linear regression model of cost of heat supply systems suggested that linear combination that could cover the highest possible potion of variance in response was formed by gross external floor area (GEFA), number of storeys above ground (NoS_{ag}), developed plot area (DPA), location index (LI) and facility type (FAC). It is noted that no fitted values of reference unit quantities significantly contributed to explain any variance in cost of heat supply systems. Average floor size (AFS) that was initially hypothesized to be relevant for the model dropped from the final model due AFS and DPA tended to explain the same portion of variance in the response variable. Hence developed plot area presented higher statistical significance, it was retained in the model. The final model accounted for 77% of variance in cost of heat supply systems (Table 5.7). Coefficients of indicator variables of facility type were estimated as $\beta_c^{k420} = (1.17581, 0.09965, 1.24328, 1.73428)$ for education, industrial, office and other facility types, respectively. The result suggested that average expected cost of heat supply systems for residential projects were the smallest when compared to the other types of facilities. Residentials are followed by industrial, educational, office and other facility types, respectively, after adjusted with the significant drivers of *k420* model.

According to the analysis conducted, one could see in Table 5.7, significant drivers of cost of air treatment systems was determined as gross building volume (GBV), average floor size (AFS), fitted values of external (m330_{fit}) and internal walls (m340_{fit}), client (CT) and facility type (FAC). Initially assumed predictors of number of storeys (NoS), average storey height (ASH), topography class (TOP) and fit of excavation volume (m310_{fit}) became redundant and dropped from the final model. It could be argued that excavation volume and topography class showed no practical relevance; whereas the variance can be explained by number of storey and average storey height, could also be explained by other macro determinants such as gross building volume and average floor size. It could also be argued that air treatment was associated with and shall be explained by volume information rather than floor area. Results of the analysis concreted this argument as parameter of gross building volume became significant whereas gross external floor area could not fulfill significance level of 10% through stepwise regression. The final model accounted for 54% of variance in the response, which could be regarded as *less than ideal* for predictive aims. Coefficients of indicator variables of facility type were estimated as $\beta_c^{k430} = (-0.18991, -0.10463, 0.12228, 0.13767)$ for education, industrial, office and other facility types, respectively. The result suggested that average expected cost of air treatment systems for educational projects were the smallest when compared to the other types of facilities. Educationals were followed by industrial, residential, office and other facility types, respectively, after adjusted with the significant drivers of *k430* model. Lastly, coefficient of public buildings was estimated as $\beta_p^{k430} = (0.16547)$ which indicates higher value of expected average cost of sewerage air treatment systems when compared to private developments.

Cost of power installations (k440) model suggested gross external floor area (GEFA), gross building volume (GBV), plot area (PA), average storey height (ASH), average floor size (AFS), topography class (TOP), market conditions (MC), facility (FAC) and client type (CT), along with fitted values of external walls (m330_{fit}), internal walls (m340_{fit}), floors and ceilings (m350_{fit}) were the best subset of variables that could explain the highest possible amount of variance in the model response. While the statistics favored the results, we fail to explain the motivation stands for practical relevance why some variables was retained in the model such as topography class. Yet, it could be argued due to high level of multi collinearity, particularly for this model, it was arduous to interpret the contributions of individual of the model. The model was accepted as it was developed for the reason that it fulfilled its primary aim of prediction in the test sample when compared to other possible combinations. The full model accounted for 83% of variation in the response variable (Table 5.7). Coefficients of indicator variables of facility type were estimated as $\beta_c^{k440} = (0.21400, 0.00207, 0.29490, 0.22052)$ for education, industrial, office and other facility types, respectively. The result suggested that average expected cost of power installation systems for residential projects were the smallest when compared to the other types of facilities. Residentials were followed by industrial, education, others and office facility types, respectively, after adjusted with the significant drivers of *k440* model. Lastly, coefficient of public buildings was estimated as $\beta_p^{k440} = (0.08285)$ which indicates slightly higher value of expected average cost of power installations when compared to private developments.

Cost of telecommunications and other communication systems (k450) indicated the cost of individual systems comprise the associated distributors, cables and wiring. The regression analysis suggested the final model, consisted of gross external floor area (GEFA), number of storeys (NoS), developed plot area (DPA), facility (FAC) and client type (CT), along with fitted value of floors and ceilings area ($m350_{fit}$), accounted for 67% variance in the model response (k450). Determination of coefficient indicated a fit which is presumably *less than ideal* for predictive aims. Like in majority of the models associated with cost of services, gross external floor area covered the largest portion of variance when compared to other significant factors. Due to multi collinearity observed in between gross floor area, developed plot area and fit of floors and ceilings area, parameters of DPA and $m350_{fit}$ were estimated as negative. Topography class was initially pointed by stepwise procedure, yet directly dropped due it showed of no practical relevance. The explanatory power of gross external floor area was significantly dominant in this cost group, which made not only other macro determinants but also fitted values of reference units, that were initially hypothesized, redundant for the final model. Coefficients of indicator variables of facility type were estimated as $\beta_c^{k450} = (0.04568, -0.00762, 0.11827, 0.06115)$ for education, industrial, office and other facility types, respectively. The result suggested that average expected cost of telecommunication and other communication systems for industrial projects were the smallest when compared to the other types of facilities. Industrials were followed by residential, education, others and office facility types, respectively, after adjusted with the significant drivers of *k450* model. Lastly, coefficient of public buildings was estimated as $\beta_p^{k450} = (0.04405)$ which indicated slightly higher value of expected average cost of power installations when compared to private developments.

Considerable number of observations in the sample indicated cost of transport systems (k460), which corresponded to the cost of lifts, escalators and etc., as *zero*. This was to say, many projects do not require transport systems and therefore spending that could be classified under this cost group was zero. The best linear combination that explained the highest possible amount of variance in the response variable, was given in Table 5.7, was formed by average floor size (AFS), fitted values of external ($m330_{fit}$) and internal walls ($m340_{fit}$), facility (FAC) and client types (CT). Parameters of gross external floor area and fitted value of foundation area showed no significance contribution to the model at 5% significance level. It could be argued that both of these variables became redundant hence they present strong linear dependency with fitted values of reference unit quantities that were retained in the model as well as average floor size. Practical experience also suggested number of storeys might provide a significant portion of variance in the response. Yet, it could also be argued that facility type (FAC) distinguishes in between projects having different number of storeys, which made the variable redundant. The final model explained 64% of the variance in the response variable. Coefficients of indicator variables of facility type were estimated as $\beta_c^{k460} = (-5.0978, -6.6554, 0.8741, 2.8006)$ for education, industrial, office and other facility types, respectively. The result suggested that average expected cost of transport systems for industrial projects were the smallest when compared to the other types of facilities. Industrials were followed by education, residential, office and other facility types, respectively, after adjusted with the significant drivers of *k460* model. Lastly, coefficient of public buildings was estimated as $\beta_p^{k460} = (-7.1680)$ which indicated lower value of expected average cost of power installations when compared to private developments. It shall be underlined, among all developed models, which retained client type as a predictor, *k460* was the only one in which lower value under public administration was observed.

Similar to *k460*, spending on cost of function-related equipment and fitments (k470), identified as none (zero) for majority of the objects in the sample. The cost group contained a wide variety of cost elements, such as cost of kitchen fitments, fire-fighting installations, disposal facilities, swimming baths equipments, made it arduous even to hypothesize a relevant driver apart from gross external floor area. The final model accounted for only 33% of variance in the response variable. This value was not satisfactory for predictive aims. One could see in Table 5.7 that the final consists of gross external floor area (GEFA), number of storey above ground (NoS_{ag}) and fitted value of internal walls area ($m340_{fit}$). Majority of variables, which were retained in the initial model, were dropped from the final model. It could be argued this is a consequence of collinearity. Hence, zero values in the sample were excluded to model *k470* the sample size shrinks dramatically. Multi collinearity influences higher standard errors and therefore lower t scores of estimated parameters. This led to high number of statistically insignificant predictors. Yet, relatively low level of coefficient of determination suggest that available information was not enough to explain enough variance in cost of function related equipment and fitments to gain accurate predictions.

Cost of building automation systems (k480) was another cost group where majority were valued as zero. The best linear combination was determined as gross external floor area (GEFA), plot area (PA), location index (LI), market conditions (MC). The final model's coefficient of determination was 89%which indicated the portion of explained variance in the response. It can be argued that especially facility type (FAC), which showed practical relevance for k480, became insignificant due to insufficient number of observations in the sample. Parameter of location index (LI) was estimated as negative. The reason might be that building automation was relatively luxurious for an average project in the sample's time-frame (1973-2010). It was likely that for objects that were located in the country side, transportation and installation of these systems could be more expensive than the projects in the urban areas.

The last cost group of building services was k490 refers to cost of other services related work, such as final disposal of materials and scaffolding. Only 28 objects had information in this cost group. A simple linear regression model was determined. The model contains gross external floor area (GEFA) as the predictor and explains only 16% of variance in the response.

In conclusion, linear models were developed for each of the 17 second level cost groups of structure. It could be argued that revealed information in the first part of the analysis (Section 5.1) made significant contribution especially for cost groups of construction works. These quantities of building elements, which was regarded as *reference unit quantities*, were able to cover significant portions of variance in responses so that gross external floor area became redundant, except one case cost of internal walls (k340). This provides a hint that estimation of reference unit quantities *may* not be for nothing.

Prediction accuracy indicators, namely root mean square error (RMSE) and mean absolute percentage error (MAPE), for developed linear models were summarized in Table 5.8. The indicators were calculated using the test sample, consisted of 49 objects. The results indicated 4 of the 6 *major* cost groups, which were directly associated with reference unit quantities, reached to a prediction error less than 30%. This was under the critical level suggested by Ruf (2003), and used as a benchmark through interpretation of prediction accuracy in cost prediction. In fact, it is the only argument regarding prediction accuracy in Germany. It would be underlined that the weight of these *major* cost groups in overall cost of structure is relatively high. Prediction accuracy for cost of foundation (k320) was *less than ideal*, considering

Table 5.8: Prediction performance of linear models for second level cost group of structure

Model	Performance measure		
	RMSE	MAPE	n^*
k310	125,248	50.85%	6
k320	144,851	32.89%	3
k330	566,417	28.07%	2
k340	207,930	27.75%	4
k350	421,106	28.27%	4
k360	190,931	22.81%	3
k370	143,073	59.15%	23
k390	145,768	52.39%	15
k410	166,454	41.03%	6
k420	103,332	32.90%	5
k430	212,354	66.41%	20
k440	188,019	37.46%	4
k450	197,463	52.48%	14
k460	62,163	53.20%	28
k470	688,757	79.49%	33
k480	1,033,817	77.32%	47
k490	36,597	78.29%	46

Note: "n^*" is number of outliers and/or zero values of responses in the test group

its relatively high value of coefficient of determination. Low prediction accuracy of the cost of excavation, on the other hand, was expected, hence associated model only covered 58% of variance. In addition, number of detected outliers for excavation cost prediction in the test sample was relatively high when compared to the rest five of the six major cost groups.

Three cost models of services, sewerage, water and gas systems (k410), heat supply systems (k420), and power installations (k440), reached to prediction error approximately between 30% and 40%. These are the cost groups which constitutes the major portion in overall cost of services. Prediction errors of the rest 6 cost groups of services varied between 50% to 80%, indicating a *poor* prediction ability. It was argued, this figure was reasoned by neither limitations implicit to linear model specifications nor possible mistakes inherit to modeling process. The reason was, crucial information, which can account for higher portion of variance to enhance predictions in these cost groups, was simply lacking. Yet, it would be emphasized the primary aim of this study was set to provide estimates for aggregate cost of structure (k300+k400). Error in individual cost groups were important and shall be interpreted but not vital for the ultimate aim of the research. Lastly, one could see in Table 5.8 that some cost groups had high number of projects excluded from performance indicator computation. In fact, test sample, like train sample, also consisted of zero values. Thus, excluded objects were mostly not outliers but *zero* values in the test sample that were assumed to be known by the architect.

As in reference unit quantity modeling, the next step was to implement artificial neural network (ANN) method to model second level cost groups of structure. Identical heuristic approaches were adopted to select type and architecture of the ANN models as mentioned in Section 5.1. Table 5.9 and 5.10 summarized typology and prediction accuracy indicators of ANN models developed for second level cost groups of structure, construction works, and services, respectively.

The neural net model which minimizes the root mean square error (RMSE) of test sample's cost of excavation (k310) was determined as a General Regression Neural (GRN) Net. Detecting

Table 5.9: Architecture and prediction performance of ANN models for second level cost groups of structure, construction works

ANN architecture and	Model							
performance parameters	k310	k320	k330	k340	k350	k360	k370*	k390
GRN Net	■						■	
MLF Net		■	■	■	■	■		■
No. nodes in hidden layer	-	4	7	15	20	4	-	3
RMSE of best configuration	122,710	139,097	345,604	215,315	317,784	186,998	147,974	124,632
MAPE of best configuration	61.99%	30.60%	28.41%	26.20%	44.50%	29.03%	118.71%	63.22%
n^*	7	2	4	4	5	4	23	5

7 objects as outliers in the test sample, mean absolute percentage error (MAPE) of the best configuration (GRN Net) was computed as 62%. This was significantly larger than MAPE of test sample when LRM model was employed to predict the value of interest, in this case cost of excavation. Considering RMSE as the measure of accuracy, ANN model could be slightly favored over linear regression model (LRM). Yet, substantial difference observed between MAPE values as well as smaller number of outliers detected in the test sample, it could be argued that LRM provided superior results to ANN when prediction of excavation cost (k310) is the aim.

The analysis revealed that the MLF Net, with 4 nodes in its hidden layer, minimized RMSE of predictions of foundation cost by the test sample. The ANN detected 2 observation as outliers. This was one observation less when LRM was considered for predicting test sample's cost of foundation. Both measures of prediction accuracy that were used in this study, MAPE and RMSE, did not show substantial differences in between different methods of analysis. (Table 5.8 and 5.9).

One could see in Table 5.9, an MLF type Net with 7 nodes in its hidden layer was selected due to the smallest RMSE value calculated for prediction of test sample's cost of external walls (k330). When ANN model was considered, 4 objects were detected as outliers. This was 2 objects more than LRM results. When MAPE was considered as measure of accuracy two method of analysis did not present disparity. Yet, when comparison was performed on the basis of RMSE, ANN model present superior prediction performance.

Cost of internal walls (k340), by the test sample, could best be predicted by an ANN model which was an MLF type, having 15 nodes in its hidden layer. Generally, a neural network tend to over-train as the number of neurons in the hidden layer increases. However, it was observed that RMSE of prediction by the test sample tend to decrease continuously until 15 neurons in the hidden layer has reached. From that point, an increasing trend observed which reveal a clear indication of over-train. MAPE of ANN model provided a slight improvement to LRM. Reversely, RMSE of LRM was slightly smaller than ANN model. Both models detected 4 observation as outliers in the test sample.

Prediction error of test sample could best be minimized by an MLF Net having 20 neurons in its hidden layer for cost of floors and ceilings (k350). ANN model detected 5 objects as outliers in the test sample. This number reduced to 4 when LRM was considered. It was evident by the results that MAPE of LRM was significantly smaller than ANN model, having the values of 28.7% and 44.50%, respectively (Table 5.8 and 5.9). Reversely, RMSE of ANN model, to predict unseen data for cost of floors and ceilings, was substantially smaller than its counterpart, LRM.

Employing the same inputs and data set in which outliers were excluded for training, roof cost (k360) of the test sample could best be predicted by an MLF Net with 4 neurons in hidden layer. ANN model detected 4 objects as outliers whereas this number reduced to 3 when LRM was considered (Table 5.8 and 5.9). In addition, the prediction accuracy was increased by 7% when LRM was considered over ANN model. However, no substantial differences was observed between associated (k360) ANN and LRM when RMSE was considered as the prediction accuracy indicator.

The analysis yielded cost of structural fitments (k370) of the test sample could be predicted with the smallest RMSE with the help of GRN Net. 23 observations, mostly *zero* valued, but not outliers were excluded from the test sample through computation of prediction accuracy indicators. While RMSE values did not show disparity in between different method of analysis, there was a significant improvement in MAPE value when LRM was considered for prediction (Table 5.8 and 5.9).

Results suggest, prediction error in the last cost group of construction works, cost of other construction related activities (k390), such as scaffolding, safety measures and final dispose of materials, could be minimized by an MLF Net with 3 nodes in its hidden layer. The number of outlying observations was detected as 5 for ANN model, whereas this number rose to 15 when LRM was considered (Table 5.8 and 5.9). When MAPE was considered as the prediction accuracy measure, LRM provided a significant improvement over ANN model. However, the number of outliers should be monitored carefully before any conclusion drawn.

ANN model which computed the lowest RMSE of prediction in cost of sewerage, water and gas systems (k410) was determined as a GRN Net. ANN model detected 5 observations as outliers, whereas the number slightly rose to 6 in case LRM was used to predict future values for cost of sewerage, water and gas systems (k410). ANN model provided superior measures of prediction accuracy, MAPE and RMSE, when compared to LRM (Table 5.8 and 5.10).

Prediction error in cost of heat supply systems (k420) could best be minimized by an MLF Net having 2 nodes in its hidden layer. The results suggested, one could expect slightly lower prediction error in cost of heat supply systems when LRM model was employed. On the other hand, RMSE value of ANN model was slightly smaller than LRM. It was noted that both models have equal number of outliers (Table 5.8 and 5.10).

According to results presented in Table 5.8 and 5.10, GRN Net could best minimize the prediction error in cost of air treatment systems (k430). Substantial differences in measure of prediction accuracies were observed in between different method of analysis, favoring LRM over ANN model. Number of outliers, on the other hand, were detected in the same levels.

Findings suggested, prediction error in cost of power installations (k440) could best be minimized by a GRN Net. Measures of prediction accuracies were close to each other, presenting no substantial difference in between method of analysis (Table 5.8 and 5.10). Detected number of outliers in the test sample was also identical.

Results in Table 5.8 and 5.10 demonstrated, RMSE error in cost of telecommunication and other communication systems (k450) was computed as the lowest when an MLF Net with 4 neurons in hidden layer was considered. 12 objects were excluded from the test sample through computation of prediction indicators by ANN model. This number rose to 14 when LRM was considered. Findings suggests, LRM offers a substantial improvement in prediction accuracy than ANN model, when MAPE was considered as the indicator. On the other hand, a slight different was present in terms of RMSE in between alternative method of analysis.

Table 5.10: Architecture and prediction performance of ANN models for second level cost groups of structure, services

ANN architecture and performance parameters	Model								
	k410	k420	k430	k440	k450	k460	k470	k480	k490
GRN Net	■		■	■		■	■		■
MLF Net		■			■			■	
No. nodes in hidden layer	-	2	-	-	4	-		8	
RMSE of best configuration	57,034	97,020	418,372	208,002	146,097	59,219	647,912	745,270	22,594
MAPE of best configuration	36.60%	35.03%	201.58%	38.97%	70.27%	51.20%	155.93%	72.68%	39.12%
n^*	5	5	19	4	12	31	33	47	46

According to findings, RMSE in cost of transport systems by the test sample could best be minimized by a GRN Net. The number of excluded observations from the test sample due zeros or extreme values (outliers) was 31 when ANN model was employed to predict the value of interest in the test sample. The number shrunk to 28 when LRM was utilized (Table 5.8 and 5.10). Both measures of prediction accuracies, MAPE and RMSE, did not present substantial alterations when alternative models were considered.

Results also revealed, prediction error in cost of function related equipment and fitments could best be minimized by the GRN Net. Excluding 33 observations form the test sample, because the values indicated as zero or outliers, for both method of analysis, prediction indicator of MAPE for LRM model were computed significantly smaller than ANN model. However, a very slight difference observed when RMSE was considered as the measure of accuracy (Table 5.8 and 5.10).

Having so many zero values in the test sample for cost of building automation (k480) and other services related work (k490), it was hard to discuss on measures of prediction accuracy. This was why, a measure of central tendency would be valid in case the sample size reached to a certain level. In this regard, *mean* absolute percentage errors (MAPE) of k480 and k490, which could only be calculated on the basis of 2 and 3 objects in the test sample, respectively, did not provide information to generalize. It could also be argued that pattern recognition capabilities of ANN models, which were trained over smaller number of objects, might not be consistent.

In consistent with the findings presented in Table 5.8, 5.9 and 5.10, it could be argued that linear regression models (LRM) reached to smaller prediction errors than artificial neural network (ANN) models in majority of the second level cost groups of structure. Besides, the number of outliers detected did not show significant disparity in majority of the cost groups. Moreover, one could see that deviation in mean absolute percentage errors of linear regression models of second level cost groups was significantly less than its counterpart, artificial neural network models. Accordingly, the range of MAPE in the test sample was computed between 22% and 78%, when LRMs were considered. The figure changed dramatically in case of ANN models and MAPEs in the test sample were calculated in the range of 26% to 201%. This suggested that ANN models were weaker based on not only their prediction ability but their reliability of pattern recognition for the problem of interest, in this case prediction of second level cost groups of structure. This led to a main conclusion, in our case, the problem did not lie in the form of function (linear or non linear) but more lacking information at this stage of development that could account for more variance in the value of interest.

5.2.2 First level cost groups of structure

Developing predictive models of first level cost groups of structure was conducted in two sections according to the cost prediction frameworks that were illustrated in Figure 1.4. Framework #2, #3 and #4 was directly associated with prediction of first level cost groups, namely cost of construction works (k300) and cost of services (k400). Framework #2 corresponded to multi-way forecast approach in which fitted (predicted) values of reference unit quantities, such as external walls area, were fed back to systems inputs, aiming to predict first level cost groups (k300 and k400) separately. Framework #3 and #4, on the other hand, associated with one-step ahead forecast type, in which only the initial macro determinants, that were assumed to be available to the architect at HOAI basic evaluation stage, were utilized to predict unseen data. Framework #3 aimed to estimate fitted values of k300 and k400 separately whereas Framework #4 targeted predicting value of k300+k400, directly. The findings were demonstrated on the basis of forecast approach, either multi-way or one-step ahead.

Table 5.11: Candidate predictors of linear regression models for the first level cost groups of structure via multi-way approach

Var.	Initial model in which the factor is included		
	k300	k400	k300+k400
GEFA	■	■	■
GBV	■	■	■
NoS		■	
NoS_{ag}	■		■
NoS_{bg}	■		
PA			
DPA	■	■	■
ASH	■	■	■
AFS	■	■	■
LI	■		
FAC	■	■	■
GC			
SA	■	■	
TOP			■
MC	■		■
CT	■	■	■
$m310_{fit}$			■
$m320_{fit}$	■		■
$m330_{fit}$	■	■	■
$m340_{fit}$	■		■
$m350_{fit}$			■
$m360_{fit}$	■	■	■

5.2.2.1 Multi-way forecast

As mentioned earlier, multi-way forecast employed, not only macro determinants available to an architect at HOAI basic evaluation stage, but also fitted values of reference unit quantities predicted in Section 5.1, as inputs. The same training and testing data sets, consisted of 475 and 49, respectively, were employed as in Subsection 5.2.1.

Identical modeling methods and assumptions were considered. Accordingly, the initial step were to determine candidate predictors that would be retained in the initial models. Stepwise regression method was utilized for variable selection procedure. The results were outlined in Table 5.11.The analysis suggests, gross external floor area (GEFA), gross building volume (GBV), developed plot area (DPA), average storey height (ASH), average floor size (AFS), facility (FAC) and client type (CT), fitted values of external walls area ($m330_{fit}$) and roof area ($m360_{fit}$) were retained in all initial models, namely cost of construction works (k300), services (k400), and structure (k300+k400). In addition, number of storeys above ground (NoS_{ag}), site access (SA), market conditions (MC), fitted values of foundation area ($m320_{fit}$), and internal walls area ($m340_{fit}$) were initially appeared in two out of three models. Special attention was paid to the number of appearance in the initial cost models of the fitted values of reference unit quantities due to adopted approach, multi-way forecast. Accordingly, it could be seen in Table 5.11 that all six reference unit quantities were appeared at least in one initial model.

Joint results of ordinary least square parameter estimation for first level cost groups of structure models were demonstrated in Table 5.12. The findings revealed linear combination of gross external floor area (GEFA), gross building volume (GBV), number of storeys above

(NoS_{ag}), and below ground (NoS_{bg}), developed plot area (DPA), average storey height (ASH), average floor size (AFS), location index (LI), facility (FAC) and client type (CT), site access (SA), market conditions (MC), fitted values of foundation ($m320_{fit}$), external walls ($m330_{fit}$), internal walls ($m340_{fit}$) area, can account for approximately 93% of the variance in cost of construction works (k300). Only, fitted value of roof area ($m360_{fit}$) was included in the initial model and later dropped from the final one. It could be argued the portion of variance that can be explain by $m360_{fit}$, could also be explained by not only macro determinants, such as gross external floor area, but also fitted values of reference unit quantities that were retained in the final model. Undeniably, considerable amount of correlation, which was present in between gross external floor area and gross building volume (GBV) influences parameter of GBV estimated as negative. The similar issue of multi collinearity was also observed in between fitted values of reference unit quantities that were retained in the model. Accordingly, parameters of fitted values of foundation and internal walls area were estimated as also negative. In conclusion, all retained variables in the model present practical relevance and did not offend common sense as well. Coefficient of determination, which was 93%, can be identified as *more than ideal* for predictive aims. Coefficients of indicator variables of facility type were estimated as $\beta_c^{k300} = (0.56665, 0.08684, 0.49830, 0.79942)$ for education, industrial, office and other facility types, respectively. The result suggested that average expected cost of construction works for residentials were the lowest when compared to the other types of facilities. Residentials were followed by industrial, offices, education and other facility types, respectively, after adjusted with the significant drivers of *k300* model. Lastly, coefficient of public buildings was estimated as $\beta_p^{k300} = (0.23685)$, which indicated a significant increase in average expected value of cost of construction works when compared to private developments.

Table 5.12: Estimated parameters and closeness of fit for final linear regression models of first level cost groups of structure via multi-way approach

Var.	Final model response		
	$k300^{0.16}$	$ln(k400)$	$(k300+k400)^{0.22}$
GEFA	0.00081		0.0044
GBV	−0.00007	0.0001	−0.0010
NoS		0.1951	
NoS_{ag}	0.14122		
NoS_{bg}	0.27456		
PA			
DPA	0.00311	0.0021	0.0020
ASH	0.14697		
AFS	0.00088	0.0024	
LI	0.40550		
FAC	β_c^{k300}	β_c^{k400}	$\beta_c^{k300+k400}$
GC			
SA	0.08638		
TOP			−0.4644
MC	0.12555		
CT	β_p^{k300}	β_p^{k400}	$\beta_p^{k300+k400}$
$m310_{fit}$			
$m320_{fit}$	−0.00375		
$m330_{fit}$	0.00087		0.0133
$m340_{fit}$	−0.00068		−0.0071
$m350_{fit}$			
$m360_{fit}$		−0.0040	
Constant	*5.27979*	*10.9645*	*14.2073*
R^2	92.54%	83.45%	91.80%

Second column of Table 5.12 outlined the linear equation that accounted for the highest possible variance in cost of services (k400). Accordingly, one could see that the linear model which retained gross building volume (GBV), number of storeys (NoS), developed plot area (DPA), average floor size (AFS), facility (FAC) and client (CT) type, and fitted value of roof area ($m360_{fit}$) explained approximately 84% of the variance in cost of services (k400). Through discussion of the models by several second level cost groups of services, it was argued gross building volume could be a better explanatory indicator for cost of services than gross external floor area, which is pointed by DIN 277-3 as the reference unit for cost of services. Initial model of k400 included both gross external floor area and gross building volume as independent variables. Yet, contribution of gross external floor area became insignificant because the variance that could be explained by GEFA, could better or greatly be explained by GBV. Thereof, it became redundant and dropped from the final model. The result of linear model for cost of services (k400) concreted the argument previously discussed and suggests that gross building volume could be a replacement for gross external floor area as a reference unit quantity of cost of services. Another initially included variable, site access (SA) shows of no practical relevance as well as an insignificant parameter, and therefore excluded from the final model as well. Average storey height might have been isolated from the final model because a very close information could also be provided by number of storeys, which was retained in the final model. Lastly, fitted value of external walls area might have been excluded due to same principle of singularity. Coefficient of determination, which was 84%, can be identified as *ideal* for predictive aims. Coefficients of indicator variables of facility type were estimated as $\beta_c^{k400} = (0.6009, 0.0488, 0.7588, 1.4749)$ for education, industrial, office and other facility types, respectively. The result suggested that average expected cost of services for residentials were the lowest when compared to the other types of facilities. Residentials were followed by industrial, education, offices and other facility types, respectively, after adjusted with the significant drivers of *k400* model. Lastly, coefficient of public buildings was estimated as , $\beta_p^{k400} = (0.3024)$ which indicated a significant increase in average expected value of cost of services when compared to private developments.

The last model that was developed via multi-way approach was the cost of structure (k300+k400). Analysis yielded a linear equation, that contained gross external floor area (GEFA), gross building volume (GBV), developed plot area (DPA), facility (FAC) and client type (CT), topography class (TOP), fitted values of external ($m330_{fit}$) and internal ($m340_{fit}$) walls, accounted for 92% of the variance in cost of structure. Macro determinants, number of storeys above ground (NoS_{ag}), average storey height (ASH) and average floor size (AFS), which were included in the initial model, later dropped from the final model. It could be argued that macro determinants that were retained in the final model, such as gross external floor area and gross building volume, made their explanatory power redundant for the final model. This might have led statistically insignificant parameters. Similar, fitted values of reference unit quantities that were included in the initial model (Table 5.11) might have been isolated by the the ones which were retained in the final model (Table 5.12). Coefficient of determination, 92%, was *more than ideal* for predictive aims. Coefficients of indicator variables of facility type were estimated as $\beta_c^{k300+k400} = (0.4505, -1.4567, 0.3703, 1.0478)$ for education, industrial, office and other facility types, respectively. The result suggested that average expected cost of structure for industrials were the lowest when compared to the other types of facilities. Industrials were followed by residentials, offices, educational and other facility types, respectively, after adjusted with the significant drivers of *k300+k400* model. Lastly, coefficient of public buildings was estimated as, $\beta_p^{k300+k400} = (0.8116)$ which indicated a significant increase in average expected value of cost of structure when compared to private developments.

Considering root mean square error (RMSE) and mean absolute percentage error (MAPE) as the measure of accuracies, joint results of prediction performance of first level cost groups of structure via multi-way approach were given in Table 5.13. Recall, performance measurements were calculated by employing identical test sample, contained 49 objects, which was also used for

Table 5.13: Prediction performance of linear regression models of first level cost groups of structure via multi-way approach

Performance measure	Model		
	k300	k400	k300+400
RMSE	1,128,575	796,912	1,396,923
MAPE	18.82%	34.05%	25.27%
$n*$	1	3	1

Note: "n*" stands for number of outliers in the test sample

measuring prediction performance of second level cost groups of structure models. In Table 5.13, one could see that prediction accuracy of cost of construction works (k300) was significantly computed below critical value of cost estimation, 30%. Only one observation, out of 49, in the test sample did not follow the pattern recognized by the linear model of k300. Secondly, prediction accuracy of cost of services model (k400), which presented relatively a lower fit to the training data, was calculated as 34%. This was above the critical value indicated in the literature by Ruf (2003). In addition, the model detected 3 observations as outliers. The results showed of no surprise for us. It was evident by the prediction accuracy of second level cost groups of services model (Table 5.8) that the crucial information was lacking to provide accurate predictions for cost of services, even after revealing reference unit quantities. Lastly, prediction accuracy of cost of structure (k300+k400) linear model was computed as 25%, which was below the threshold value of early cost estimation.

Table 5.14: Architecture and prediction performance of ANN models for first levels cost groups of structure via multi-way approach

ANN architecture and performance parameters	Model		
	k300	k400	k300+k400
GRN Net	■	■	■
MLF Net			
No. nodes in hidden layer	-	-	-
RMSE of best configuration	1,038,503	699,891	1,408,560
MAPE of best configuration	23.65%	39.73%	23.36%
$n*$	-	3	-

Following, artificial neural network (ANN) models were developed. Repeatedly, ANN models employed identical training and test data sets to enable a fair comparison. Input nodes of ANN models were those of variables which made significant contribution to associated linear regression models. Finally, identical heuristic approaches were adopted as mentioned in Section 5.1 through determination of network typology.

Collective findings of ANN models were presented in Table 5.14. The analysis revealed that GRN Nets minimized RMSE of predictions of cost of construction works (k300), services (k400) and structure (k300+400) by the test sample. Considering mean absolute percentage error (MAPE) as the measure of prediction accuracy, linear regression model (LRM) of cost of construction work (k300) provided approximately 5% improvement over its counterpart, ANN model. Yet, ANN model's RMSE value was slightly lower than LRM for the same model of interest, construction works. It would be noticed that ANN model detected no outliers in he test sample. Similar to the model of construction works (k300), ANN model for cost of services (k400) showed a lower prediction accuracy by 5% when compared to LRM (Table 5.13 and 5.14). Recall both method of analysis detected 3 objects as outliers. Finally, ANN model for cost of structure (k300+400) did not present a substantial difference in MAPE to LRM model. Besides the other indicator of prediction accuracy, RMSE, was also calculated as very close values, showed no significant disparity in between alternative method of analysis (Table 5.13 and 5.14).

In conclusion, pairwise comparison results between methods of analysis, LRM and ANN, revealed that prediction accuracy of LRMs were superior to ANN models under assumed conditions. The result suggested, neither functional form of pre assumed functional relationship in between variables, nor model assumptions to be fulfilled, was the main issue in predicting construction costs at the conceptual stage of developments. The reason was, presumably, lacking of crucial information which may account for additional portion of variance in the value of interests that could not already be covered by the information available.

5.2.2.2 One-step ahead forecast

The last step of cost estimation was the application of one-step ahead forecast approach. In one-step ahead approach possible model inputs were restricted to those which were available to the architect at the HOAI basic evaluation stage. Thereof, fitted values of reference unit quantities were not employed through modeling. Two cost estimation frameworks were associated with the one-step ahead approach. Framework #3 which predicted cost of construction works (k300) and services (k400), separately and later aggregated them to compute cost of structure; and Framework #4 which directly targeted prediction of cost of structure (k300+400) with the restricted information.

The difference between multi-way an one-step ahead approaches was not only the variables could be used as inputs but also the size of training data. Through multi-way approaches initial training data was filtered due to missing values in fits of reference unit quantities. Consequently, multi-way approach employed 475 objects as the training sample. However, one-step ahead approach did require the information that was derived by the prediction of reference unit quantities. Thereof, the full training data, consists of 600 objects, could be employed through modeling. This might increase the capacity for pattern recognition which could be regarded as an advantage when compared to multi-way approach. While training data was changed, identical test sample of 49 objects were held constant as it was used for multi-way approach. The motivation was to form a basis for unbiased and fair comparisons in between different frameworks of cost estimation. This comparison was discussed in Chapter 6 in detail.

Stepwise regression at 10% significance level was employed for variable selection procedure. The variable that were retained in the initial linear models of cost of construction works (k300), services (k400), and structure (k300+k400) via one step ahead approach was given in Table 5.15. It could be tracked in Table 5.15 that, gross external floor area (GEFA), plot area (PA), developed plot area (DPA), average floor size (AFS), facility (FAC) and client type (CT) and site access (SA) were included in all initial models. Following, gross building volume (GBV), number of storeys (NoS), average storey height (ASH) and market conditions (MC) appeared at least in two models.

Ordinary least square parameter estimation results of first level cost groups of structure linear models via one-way approach were presented in Table 5.16. The results revealed the linear combination of gross external floor area (GEFA), number of storeys above ground (NoS_{ag}), plot area (PA), average floor size (AFS), site access (SA), market conditions (MC), facility (FAC) and client (CT) type could best explain the highest possible amount of variance in cost of construction works (k300). It would be underlined restriction of input variables significantly reduced chance of observing issues related to collinearity. In fact, all parameters in cost of construction works model were estimated as positive, like one might expect by his common sense.

Table 5.15: Candidate predictors of linear regression models for the first level cost groups of structure via one-step ahead approach

Var.	Initial model in which the factor is included		
	k300	k400	k300+k400
GEFA	■	■	■
GBV		■	■
NoS		■	■
NoS_{ag}	■		
NoS_{bg}			
PA	■	■	■
DPA	■	■	■
ASH	■	■	
AFS	■	■	■
LI			
FAC	■	■	■
GC			
SA	■	■	■
TOP			
MC	■		■
CT	■	■	■

The parameters of two variables, developed plot area and average storey height, which might influence multi-collinearity were estimated as statistically insignificant. Therefore, developed plot area (DPA) and average storey height (ASH) that were included in the initial model, were not retained in the final model. It could be argued, information that could be revealed by average storey height might have been isolated by combination of number of storeys above ground and gross external floor area. Developed plot area, on the other hand, showed a strong linear pattern with average floor size which might have made the variable redundant for the final model, too. Resulted regression equation could account for 90% of the variance in cost of construction works. Coefficients of indicator variables of facility type were estimated as $\beta_c^{k300} = (11.0589, 3.2084, 17.8456, 17.3746)$ for education, industrial, office and other facility types, respectively. The result suggested that average expected cost of construction works for residentials were the lowest when compared to the other types of facilities. Residentials were followed by industrials, education, other and office facility types, respectively, after adjusted with the significant drivers of *k300* model. Lastly, coefficient of public buildings was estimated as, $\beta_p^{k300} = (7.5027)$ which indicates a significant increase in average expected value of cost of structure when compared to private developments.

Analysis yielded the linear model that contains gross external floor area (GEFA), gross building volume (GBV), number of storeys (NoS), plot area (PA), average storey height (ASH), average floor size (AFS), facility (FAC) and client (CT) type, could best account for the possible highest portion of the variance in cost of services (k400). There was an evidence of collinearity that the parameter of gross building volume was estimated as negative. The reason might be that both gross external floor area and gross building volume was retained in the model at the same time. Previously it was discussed that the gross building volume as the main unit for cost of services shall be considered instead of gross external floor area. The results of the linear model concreted this argument hence gross building volume could not be excluded from the model although gross floor area was retained in the model. By other words the contribution of gross building volume for the model was still significant after adjusted with gross external floor area. In addition, the final model accounted for 80% of the variance in cost of services (k400). Yet, comparing two models resulted by multi way and one step ahead approaches, relying on coefficient of determination will be biased hence training samples present disparities. It was

Table 5.16: Estimated parameters and closeness of fit for final linear regression models of first level cost groups of structure via one-step ahead approach

Var.	Model response		
	$k300^{0.33}$	$k400^{0.12}$	$(k300+k400)^{0.26}$
GEFA	0.0083	0.00017	0.0022
GBV		−0.00002	
NoS		0.13384	1.5702
NoS_{ag}	4.2497		
NoS_{bg}			
PA	0.0003	0.00001	0.0001
DPA			
ASH		0.11089	
AFS	0.0094	0.00019	0.0040
LI			
FAC	β_c^{k300}	β_c^{k400}	$\beta_c^{k300+k400}$
GC			
SA	2.3128		
TOP			
MC	4.9881		1.6294
CT	β_p^{k300}	β_p^{k400}	$\beta_p^{k300+k400}$
Constant	*43.1228*	*3.14004*	*19.8878*
R^2	90.29%	80.64%	89.46%

noted that site access (SA) was dropped from the model due practical irrelevance. It could also be argued information that could be revealed by developed plot area could also be covered by average floor size and therefore it became redundant for the final model. Coefficients of indicator variables of facility type were estimated as $\beta_c^{k400} = (0.21617, 0.06246, 0.31300, 0.37766)$ for education, industrial, office and other facility types, respectively. The result suggests that average expected cost of services for residentials were the lowest when compared to the other types of facilities. Residentials were followed by industrials, education, office and other facility types, respectively, after adjusted with the significant drivers of *k400* model. Lastly, coefficient of public buildings was estimated as $\beta_p^{k400} = (0.17652)$, which indicated a significant increase in average expected value of cost of structure when compared to private developments.

Last linear cost model developed within the context of the research was for cost of structure (k300+k400) by employing one-step ahead approach. In consistent with the results, demonstrated in Table 5.16, one could see the linear equation that contained gross external floor area (GEFA), number of storeys (NoS), plot area (PA), average floor size (AFS), market conditions (MC), facility (FAC) and client (CT) type accounts for 89% of the variance in cost of structure (k300+k400). It could be argued gross building volume, which was included in the initial model, was isolated due dominant explanatory power of gross external floor area (GEFA) for cost of structure. This might seem contradictory to what was previously argued regarding contribution of gross building volume while explaining variance in cost of services (k400). Yet, the weight of cost of services (k400), compared to construction works (k300), was significantly less in total cost of structure (k300+k400) for an average building. This might stand for the motivation why gross building volume became redundant for the final model. Coefficients of indicator variables of facility type were estimated as $\beta_c^{k300+k400} = (3.7934, 1.5632, 5.9416, 6.3717)$ for education, industrial, office and other facility types, respectively. The result suggested that average expected cost of structure for residentials were the lowest when compared to the other types of facilities. Residentials were followed by industrials, education, office and other facility types, respectively, after adjusted with the significant drivers of *k300+k400* model. Lastly, coefficient of public buildings was estimated as $\beta_p^{k300+k400} = (2.3360)$, which indicated a significant increase in average expected value of cost of structure when compared to private developments.

Table 5.17: Prediction performance of linear regression models of first level cost groups of structure via one-step ahead approach

Performance measure	Model		
	k300	k400	k300+k400
RMSE	874,392	834,626	1,492,536
MAPE	24.72%	39.56%	26.87%
n*	2	2	2

Prediction accuracy of the developed models was validated with the help of the test sample, which was identical to the one that was used through all cots estimation exercises. Collective results were presented in Table 5.17. According to the results demonstrated, all models detected 2 objects as outliers in the test sample, which contained 49 objects. The model associated with cost of construction works (k300) reached to the prediction error of 25% which was under the critical level of 30% suggested by Ruf (2003). The figure was similar for the model of cost of structure (k300+k400). When MAPE is considered as the measure of accuracy, the model (k300+k400) computes MAPE of the test sample as approximately 27%. Yet, the model of services (k400) resulted in relatively higher error rate due to lower fit of the model to the training data. Analysis revealed that any prediction might have led to an error as high as 40% in average when linear regression model of cost of services (k400) was considered for prediction aims (Table 5.17).

The next exercise, as always, was to implement artificial neural network (ANN) method to reach comparative results between alternative types of analysis. Recall, same training and test samples were utilized as linear regression analysis of one way approach. The input nodes of ANN models were those of variables that have significant regression parameters with associated response variables (Table 5.16). Heuristic methods, that were explained in detail previously, were adopted.

Table 5.18: Architecture and prediction performance of ANN models for first levels cost groups of structure via one-step ahead approach

ANN architecture and performance parameters	Model		
	k300	k400	k300+k400
GRN Net			
MLF Net	■	■	■
No. nodes in hidden layer	3	3	9
RMSE of best configuration	821,567	640,528	1,069,154
MAPE of best configuration	25.55%	38.09%	31.11%
n*	3	5	4

Joint results were presented in Table 5.18. Accordingly, MLF type Nets, with 3 neurons in their hidden layer, minimized root mean square error (RMSE) of the test sample when the aim was to predict cost of construction works (k300) and services (k400). The number of neurons in hidden layer of MLF Net was determined as 9, when consideration was prediction of cost of structure (k300+400). The ANN model reached to slightly a higher measure of prediction error (MAPE) than LRM, when the interest was to predict cost of construction works (k300). Yet the difference was not substantial, shows no significance (Table 5.18). The figure was same for cost of services (k400) as no substantial differences were observed between alternative method of analysis. However, LRM offered approximately 5% of improvement of prediction accuracy over ANN model when the intention was predicting cost of structure (k300+k400). In addition, the numbers of outliers detected were 3, 5 and 4 for the ANN models of cost of construction works (k300), services (k400), structure (k300+k400), respectively. Recall, LRM detected 2 objects as outliers for every model.

In conclusion, the results suggested pattern recognition capacity of LRM was superior to ANN models, for the population of interest, in this case German building projects, when intention was set to employ one step ahead approach. This was due smaller number of outliers detected and lower error observed through predicting unseen data. The findings also concreted the previous argument: High error rate in construction cost prediction was reasoned by neither functional form of pre assumed functional relationship in between variables, nor model assumptions to be fulfilled. The reason was, presumably, lacking of crucial information which might account for the additional variance in the value of interests that could not already be covered by the information available.

5.3 Construction Duration

Construction duration is referred to as one of the most important project strategic determinants along with cost and quality (Figure 1.1). In this context, one of the objectives of this study was set to improve accuracy of predictions as well as to develop a model which in turn may serve as a basis for further comparison.

Like cost estimation, which was demonstrated exhaustively in Section 5.2, duration estimation was conducted in accordance with proposed frameworks (Figure 1.5). Accordingly, Framework #5 corresponded to multi-way forecasting approach, in which estimated values of project relevant information, reference unit quantities and cost groups of structure, in earlier steps are used as inputs to model the value of interest, construction duration. Whereas, Framework #6 indicates one-step ahead approach, input variables were restricted to those which were available at HOAI basic evaluation stage.

One additional input variable, which had not been employed previously, was included through implementation of both frameworks. It was underlined in Chapter 4, that all related variables were adjusted to current prices by employing construction cost index. While it was possible to eliminate the effect of time on capital by using an official statistics, it was not possible to quantify the effect of time on construction duration, if any, by the same way. The only way was to include a time related variable into the data set and control its significance of contribution to the model statistically. Thereof, the variable which contained the information regarding year of construction commence (CSD) was included in the data set.

Table 5.19: Candidate predictors of construction duration models

Response	Initial model in which the factor is included	
	Multi-way approach	One-step ahead approach
CD	CSD, NoS_{bg}, ASH, $k370_{fit}$, $k310_{fit}$, $k460_{fit}$, $k340_{fit}$, GBV, $k390_{fit}$, $k330_{fit}$, $k320_{fit}$, NoS_{ag}, $k300_{fit}$, SA, FAC, CT	GEFA, CSD, NoS_{bg}, PA, GBV, ASH, GC, TOP, FAC, CT

Compilation the data set which was utilized through duration estimation via multi-way approach was performed as follows: Actual values of the reference unit quantities and second level cost groups of structure were replaced by their predicted values which were estimated by the help of models developed previously. This operation was conducted for two times, once for each method of analysis utilized in this study, namely linear regression and artificial neural networks. Later, missing values which were previously detected as the outliers through modeling was filtered from each variable. The remaining data set, which was employed in multi-way approach, comprises 423 objects. While compiling data for multi-way approach required a great deal of effort and attention, one-step ahead approach could employ the full training data set contained 600 projects. For both approaches, the test sample that was identical to cost estimation, contained 49 observations, was used. Recall, it was assumed that the architect is competent when a second level cost group requires no value, he could detect zeros *beforehand*.

Stepwise regression was employed to select the variables included in the initial models. The collective results, for both multi-way and one-step ahead approach, were presented in Table 5.19. Accordingly one can see, construction start data (CSD), number of storeys below ground (NoS_{bg}), gross building volume (GBV), average storey height (ASH), facility (FAC) and client (CT) type were included in both of the initial models. It could be underlined, initial analysis indicated seven cost groups of structure but no reference unit quantities as candidate predictors of duration model via multi-way approach. Presumably, explanatory power of reference unit quantities were isolated by the variance that could be accounted by other candidate predictors.

Table 5.20: Estimated parameters and closeness of fit for final linear regression models of construction duration

Var.	Response	
	Multi-way	One-step ahead
	$CD^{0.25}$	$CD^{0.31}$
CT	$\beta_p^{CD,mult.}$	$\beta_p^{CD,one}$
FAC	$\beta_c^{CD,mult.}$	$\beta_c^{CD,one}$
CSD	−0.0062	−0.01605
NoS_{bg}	0.2411	0.31339
$k370_{fit}$	$1.46E-06$	
$k460_{fit}$	$2.37E-06$	
GBV	$2.04E-05$	
GEFA		$8.82E-05$
$k390_{fit}$	$-3.28E-06$	
NoS_{ag}	0.1095	
Constant	*16.1361*	*37.5841*
R^2	50.22%	41.44%

Ordinary least square estimation results were presented in detail in Table 5.20. One could see the linear combination of construction start date (CSD), number of storeys below ground (NoS_{bg}), fitted values of cost of structural fitments ($k370_{fit}$), transport systems ($k460_{fit}$), other construction related activities ($k390_{fit}$), gross building volume (GBV), and number of storeys above ground (NoS_{ag}) could account for 50% of variation in the construction duration (CD), measured in calendar days. The results might seem odd hence the final model contains variables like cost of transport systems but not cost of internal ($k340_{fit}$) and external walls ($k330_{fit}$), which may present more practical relevance in many readers mind. Yet, the investigation of the data set provided a hint. It could not be explained by *coincidence,* all cost related variables that were retained in the final model valued of *zero* in many cases. The result suggested, if an activity that was related to one of these cost groups was conducted and therefore valued, than it was expected to observe a significant influence on construction duration. Moreover, the explanatory power of cost of construction works ($k300_{fit}$), internal walls ($k340_{fit}$), external walls ($k340_{fit}$), foundation ($k340_{fit}$), excavation volume ($k340_{fit}$), might be isolated by the variance explained by the macro determinants that were retained in the final model such as gross building volume (GBV) and average storey height (ASH). Recall, these cost values are fitted values and were initially derived by the linear functions of macro determinants available at HOAI basic evaluation stage. Lastly, null hypothesis of time has no effect on construction duration can be rejected at 5% level. The results suggested that construction duration had a time dependency. Assuming that no other variable in the final model was correlated to construction start date, the linear equation suggested the average expected construction duration reduced as the years passed. This was expected due to observed development in construction technologies within the range of investigation, 1973-2010. Coefficients of indicator variables of facility type were estimated as $\beta_c^{CD,mult.} = (0.3588, 0.0207, 0.3675, 0.5258)$ for education, industrial, office and other facility types, respectively. The result suggested that average expected construction duration of residentials were the lowest when compared to the other types of facilities. Residentials were followed by industrial, education, office and other facility types, respectively, after adjusted with the significant drivers of *construction duration* model. Lastly, coefficient of public buildings was estimated as $\beta_p^{CD,mult.} = (0.3375)$, which indicated a significant increase in average expected value of construction duration when compared to private developments.

The results revealed that using one-step ahead approach, one could explain the highest possible variance in construction duration by a linear model consisted of construction start date (CSD), number of storeys below ground (NoS_{bg}), gross external floor area (GEFA), facility (FAC) and client (CT) type. The final model suggested portion of variance that could be explained by gross building volume (GBV), plot area (PA), average storey height (ASH), ground conditions (GC) and topography class (TOP), could also be explained by the macro determinants retained in the final model. Therefore, these variables became redundant, and excluded from the final model. The final model suggested construction duration had time dependency, like in multi-way approach model. In fact, estimated parameter of construction start date (CSD) was significant at 5% level. Final form of the model could account 41% of the variance in construction duration , measured in calendar days. Coefficients of indicator variables of facility type were estimated as $\beta_c^{CD,one} = (0.5513, -0.1526, 0.3043, 0.6537)$ for education, industrial, office and other facility types, respectively. The result suggests that average expected construction duration of industrials were the lowest when compared to the other types of facilities. Industrials were followed by residential, office, education and other facility types, respectively, after adjusted with the significant drivers of *construction duration* model developed via one-step ahead approach. The results presented disparity when compared to multi-way approach in which residentials were estimated to have the lowest average construction duration followed by industrials, educations, offices and others. It could be argued multi-way approach was capable of accounting for a higher portion of variance and the results suggested by multi-way approach would be more precise than one-step ahead approach. Lastly, coefficient of public buildings was estimated as $\beta_p^{CD,one} = (0.5580)$, which indicated a significant increase in average expected value of construction duration when compared to private developments. This was parallel to the findings of the model by multi-way approach.

Table 5.21: Prediction performance of linear regression models of construction duration

Performance measure	Model of CD	
	Multi-way	One-step ahead
RMSE	214.28	249.59
MAPE	30.30%	33.64%
n^*	5	4

Prediction performance measurements, root mean square error (RMSE) and mean absolute percentage error (MAPE), of linear models with respect to adopted forecasting approach was given in Table 5.21. It was evident by the performance indicators, computed on identical test sample, multi-way approach offered an improvement in prediction accuracy compared to one-step ahead approach. Yet, the difference was not substantial. It was also noted number of objects detected as outliers in the test sample is 5 and 4 by multi-way and one-step ahead approach, respectively.

Table 5.22: Architecture and prediction performance of ANN models for construction duration

ANN architecture and performance parameters	Model of CD	
	Multi-way	One-step ahead
GRN Net	■	
MLF Net		■
No. nodes in hidden layer	-	4
RMSE of best configuration	215.78	208.64
MAPE of best configuration	29.32%	28.14%
n*	5	5

Finally, artificial neural networks (ANN) were adopted to develop last two models within the content of this research. Same modeling principles were applied as before. This involved in employing the identical training and test samples to enable a fair comparison, and using significant drivers of the final linear models as the input nodes. Determination of type and architecture of the ANN models as well as addressing elimination problem of over-training, were also based on the same heuristic approached, explained exhaustively in Section 5.1. The collective results were presented in Table 5.22

Results of analysis yielded General Regression Neural (GRN) type Net was able to compute the lowest RMSE in test sample by *multi-way approach* and therefore selected to predict future values of construction duration for unseen data. The best ANN model calculated MAPE of test sample as 29%, which offered a very slight improvement over corresponding linear regression model (LRM). Similar, RMSE and number of outliers detected in the test sample shows no substantial differences in between method of analysis (Table 5.21 and 5.22).

Results of analysis showed a Multi Layer Feedforward (MLF) type Net with 4 neurons in its hidden layer minimized RMSE in test sample when one-step ahead approach was considered. Thereof, it was selected to be used for predictive aims. MAPE of test sample was computed by ANN model, as 28%. This indicates 5% improvement to corresponding LRM. RMSE of the test sample, on the other hand, was calculated approximately 20% less when compared to LRM. However it was noted that ANN model detects an extra object as an outlier.

One important outcome of this section was to reveal time dependency of construction duration. To our knowledge, this have not been addressed in relevant studies, before. For both forecast approaches, estimated parameter of construction start date, measured in years, was statistically significant at 5% level.

Results of linear models indicated prediction performance of multi-way approach was slightly superior to one-step ahead approach. However, the figure changed when ANN was considered as the method of analysis. Prediction performance of ANN models were very close each other. Moreover, neither LRM nor ANN models was capable of providing accurate predictions. The prediction performance of the models could be regarded as *fair*. Finally, the results suggested expansion of context of the argument, which was stated subsequent to cost estimation models. Repeatedly, we argued, high error rates in prediction of project determinants were reasoned by neither functional form of pre assumed functional relationship in between variables, nor model assumptions to be fulfilled. The reason was, presumably, lacking of crucial information which might account for the additional variance in the value of interests that could not already be covered by the information available.

6 Comparison of the Frameworks

Chapter 6 aimed to demonstrate comparison of the cost and duration estimation frameworks which were offered in Subsection 1.3.3, Figure 1.4 and 1.5. These frameworks formed not only the main flow of information through conducted analysis, but also the main hypotheses to be tested. Therefore, they were entitled to be reported in a separate chapter.

Discussions regarding individual models of reference unit quantities, second level cost groups of structure, first level cost groups of structure and construction duration, were performed exhaustively in Chapter 5. In addition, Chapter 5 reported comparisons between alternative method of analysis, linear regression and artificial neural networks, depending on measures of prediction accuracies for each individual model developed within the content of the research. Therefore, the scope of Chapter 6 was strictly limited to comparisons between offered frameworks.

6.1 Cost Estimation Frameworks

Repeatedly, four frameworks were offered to estimate cost of structure. Framework #1, initiated the process with estimation of reference unit quantities by using macro project determinants that were assumed to be available to an architect at the HOAI basic evaluation stage. Actual values replaced by the fitted value of reference unit quantities, which were derived from parametric functions based on linear regression and artificial neural networks. Later, these values were fed back to the system, and considered as additional information along with available macro determinants. In manuscript, this type of information flow was regarded as *multi-way* approach. Recall, multi-way approach had emerged subsequent to observed relationship between macro determinants of sampled projects and reference unit quantities through pilot analysis. Second level cost groups of structure models, such as cost of external walls, were developed using all these inputs, fitted values of reference unit quantities along with macro determinants. Aggregating predicted values of second level cost groups, one could reach to predicted cost of structure. It was hypothesized that prediction error in cost of structure could be reduced to an extent hence additional information was acquired and used as inputs. In addition, the framework provided predictions of reference unit quantities and second level cost groups of structure which in turn might be employed as benchmark values at the conceptual design phase of a building project.

Framework #2, on the other hand, adopted multi-way approach, too. It, again, commenced with developing prediction models of reference unit quantities with the assumed limited information available at HOAI basic evaluation stage. In consistent with, multi-way approach, these values were fed back to information system as inputs. The difference occurred when cost estimation was considered. Framework #2 aimed at predicting first level cost groups of structure, namely cost of construction works (k300) and services (k400). Aggregating predicted values of first level cost groups, one could reach to predicted cost of structure. One could expect an improvement in prediction accuracy due enhanced level of information used through parameter based modeling. The outcomes of Framework #2, was not as detailed as Framework #1. Second level cost groups of structure can not be estimated when Framework #2 was considered for cost estimation.

Table 6.1: Comparison of cost estimation frameworks with respect to method of analysis

Method of analysis	Framework #	Performance measurements			
		RMSE	MAPE	n*	N*
LRM	1	1,534,987	22.38%	2	33
	2	1,696,247	20.83%	2	35
	3	1,533,812	25.08%	3	30
	4	1,492,536	26.87%	2	29
ANNs	1	881,718	26.30%	-	28
	2	1,583,844	25.20%	-	33
	3	1,243,873	24.86%	3	31
	4	1,069,154	31.11%	4	26

Note: n* stands for number of outliers whereas N* indicates number of observations that have absolute percentage error of less than 30% in the test sample.

Framework #3 and #4 were developed, aimed at comparison. Their main role were to test whether any differences in prediction accuracy would occur in between multi-way and *one-step ahead* approaches. In one-step ahead approach, the input variables were restricted to those which were assumed to be available at HOAI basic evaluation stage, macro project determinants. Limited information were utilized as inputs to perform parametric estimations of first level cost groups of structure (k300 and k400) - Framework #3, and aggregated cost of structure (k300+k400) - Framework #4.

Subsequent to development of all cost models which were exhaustively demonstrated and discussed in Section 5.2, aggregated cost of structure with respect to cost estimation Frameworks, could easily be computed with respect to method of analysis. Recall, identical test sample, consists of 49 objects, was used for validation aims through cost estimation modeling. Therefore it could be argued that comparison between alternative frameworks was unbiased and fair. In addition, size of the test sample was big enough to draw meaningful statistical inferences. For instance, *mean* absolute percentage error of 5 test projects might have provided coincidental results. The size of test sample should be large enough to reveal central tendency of population of interest, in this case German building projects. Our study is significantly superior to other works which had a very limited number of validation objects for developed models.

Table 6.1 outlined collective results of comparison in between cost estimation frameworks with respect to type of analysis. Root mean square error (RMSE) and mean absolute percentage error (MAPE) of test sample were computed for comparison aims. The results revealed Framework #1, which aggregated second level cost groups to calculate predicted value of cost of structure, detected 2 observations as outliers in the test sample. Remaining 47 objects in the test sample reached to a MAPE of 22% when linear regression (LRM) method was considered.

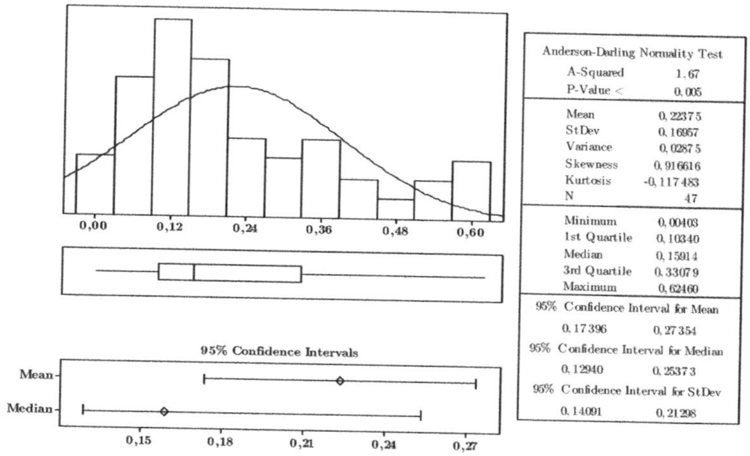

Figure 6.1: Summary statistics of test sample for absolute percentage errors of cost framework #1 via linear regression models

Absolute percentage errors (APE) of individual objects in the test sample were investigated carefully. Figure 6.1 illustrated summary statistics of APEs. One can see in Figure 6.1, APEs were positively skewed and not normally distributed. This indicated more projects in the test sample had lower error values than the MAPE value. Another indicator of central tendency, *median*, of absolute percentage errors in test sample was computed approximately 16%. This was 6% less than MAPE. It was evident that relatively smaller number of objects in the test sample reached to higher amount of error which increases the mean value substantially. Measure of dispersion in the test sample can best be given by standard deviation which was computed as 17%. Statistics of the test sample showed, one can expect APE value between 17% and 27% at confidence level of 95%. 33 objects in test sample that were below threshold value were detected; when 30% of error rate was accepted as the critical value of cost estimation practices in Germany as suggested by Ruf (2003) (Table 6.1).

When artificial neural networks (ANN) was considered as the method of analysis, Framework #1 detected no outliers. A significant reduction in root mean square error was observed when compared to LRM models. However, another indicator of prediction performance, mean absolute percentage error (MAPE), of the test sample was approximately 4% higher than LRM. Investigation of individual absolute percentage errors (APE) of the test sample, Figure 6.2, suggested a more ideal distribution. Anderson-Darling test indicated that APE terms of the test sample can be identified as normally distributed. In this case, it could be argued that mean value of APE was a clear indicator of central tendency of the test sample, when ANN was considered as the method of analysis. This argument concreted as the median of APE terms was 25%. Moreover, standard deviation of the error terms were computed as 19%. Statistics also suggest one can expect APE between 21% and 32% at 95% of confidence, when ANN models were considered for prediction (Figure 6.2). The results of analysis indicated 28 objects in the test sample reached to APE lower than 30% of threshold value, when Framework #1 via ANN models was implemented (Table 6.1).

Framework #2 adopted multi-way approach, in which fitted values of first level cost groups (k300 and k400) are aggregated to compute predicted values of cost of structure (k300+k400). Results indicated Framework #2 detected 2 outliers in the test sample when linear regression (LRM) is considered as method of analysis. MAPE of remaining 47 objects in the test sample

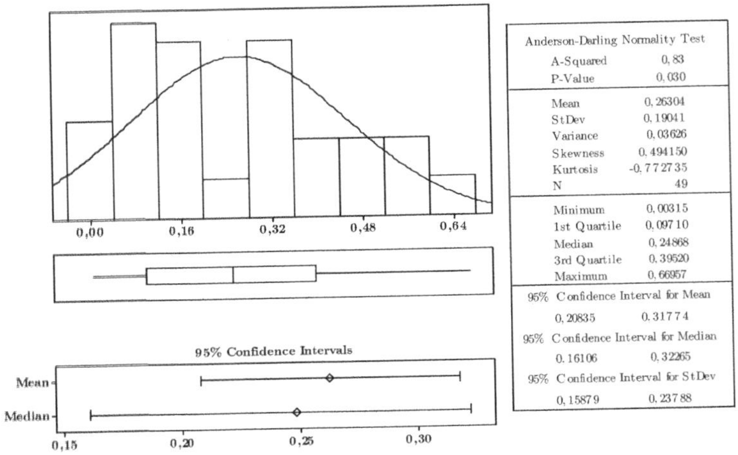

Figure 6.2: Summary statistics of test sample for absolute percentage errors of cost framework #1 via artificial neural network models

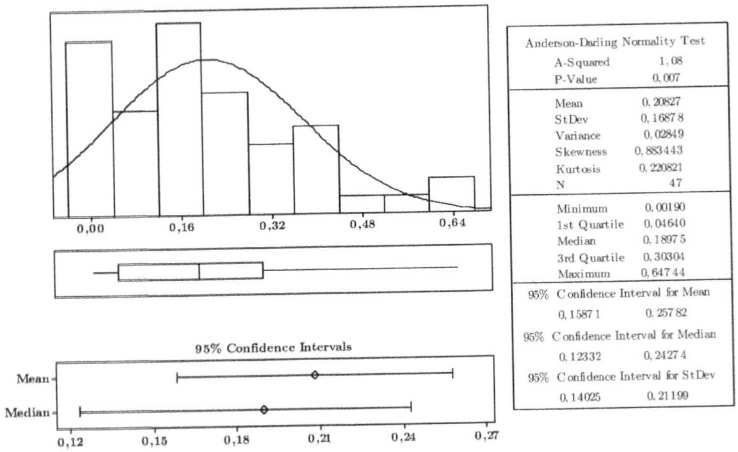

Figure 6.3: Summary statistics of test sample for absolute percentage errors of cost framework #2 via linear regression models

was calculated as 21% (Table 6.1). Individual investigation of absolute percentage error (APE) terms of the test sample, when Framework #2 via LRM method was considered for prediction, was illustrated in Figure 6.3. It was evident, APE terms of the sample was slightly skewed positively. This indicates error terms of the test sample was accumulated more in relatively smaller values. The measure of dispersion, standard deviation, was computed as 17%. Another measure of central tendency, median shall be taken into account because normality was violated. It was given as 19%. The results suggest, positive skewness of Framework #1 via LRM is greater than its counterpart, Framework #2 via LRM. In addition, 95% confidence interval of mean for APE terms of the test sample was between 16% and 26% (Figure 6.3). While the range of the interval (10%) is identical to Framework #1, the values shifted 1% in the negative direction.

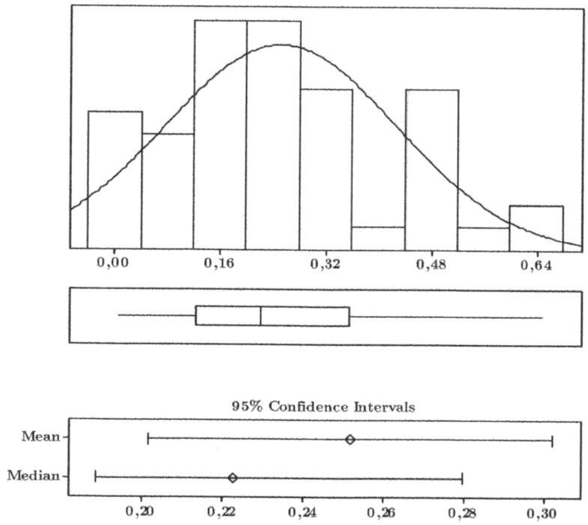

Anderson-Darling Normality Test	
A-Squared	0,83
P-Value	0,030
Mean	0,25202
StDev	0,17489
Variance	0,03059
Skewness	0,591339
Kurtosis	-0,442246
N	49
Minimum	0,00686
1st Quartile	0,12561
Median	0,22292
3rd Quartile	0,35659
Maximum	0,64841
95% Confidence Interval for Mean	
0,20179	0,30225
95% Confidence Interval for Median	
0,18872	0,27987
95% Confidence Interval for StDev	
0,14584	0,21848

Figure 6.4: Summary statistics of test sample for absolute percentage errors of cost framework #2 via artificial neural network models

As observed in Framework #1, Framework #2 also detected no outliers in the test sample when ANN models were considered as the method of prediction. Mean absolute percentage error of Framework #2 via ANN models were reported as 25%. This indicates %1 improvement when compared to Framework #1 via ANN method. However, a significant increase in root mean square error of the test sample was observed in between Framework #2 and #1 via ANN models. Investigation of individual absolute percentage error (APE) terms of the test sample was conducted and the results were outlined graphically in Figure 6.4. It was evident with Anderson Darling test that normality of error terms was not violated. Thereof, mean could be regarded as an indicator of central tendency for the APE terms resulted by the test sample. Standard deviation of APE was computed as 17%. This indicated a lower amount of dispersion in APEs when compared to Framework #1 via ANN method. With 95% confidence, one would expect that computed absolute percentage error would fall a range between 20% to 30%, when ANN method, according to flow of information in Framework #2, was applied. This indicated slightly narrower interval as well as shifted in negative direction as 1% in comparison to the interval of Framework #1 via ANN method. Last but not least, APE value of 33 objects were computed lower than the critical value of cost estimation, suggested by Ruf (2003) as 30%.

As mentioned previously, Framework #3 adopted *one-step ahead* approach. The aim was to predict cost construction works (k300) and services (k400) separately and aggregate these cost groups to determine predicted value for overall cost of structure (k300+k400). In one-step ahead approach an architect was restricted to employ the macro project determinants which are available at HOAI basic evaluation stage. Considering linear regression (LRM) as the method of analysis, 3 objects were detected as the outliers within the identical test sample that consisted of 49 objects. Graphical summary of absolute percentage errors of remaining 46 objects in the test sample was illustrated in Figure 6.5. Accordingly mean of absolute percentage errors (MAPE) was computed as 25% when LRM was implemented as method of analysis. This was higher than Framework #1 and #2. Anderson Darling statistics indicated the absolute percentage error terms did not violate normality. One should expect the true value of absolute percentage error shall lie between 20% and 30% at confidence level 95%, when LRM method according to information flow of Framework #3 was implemented.

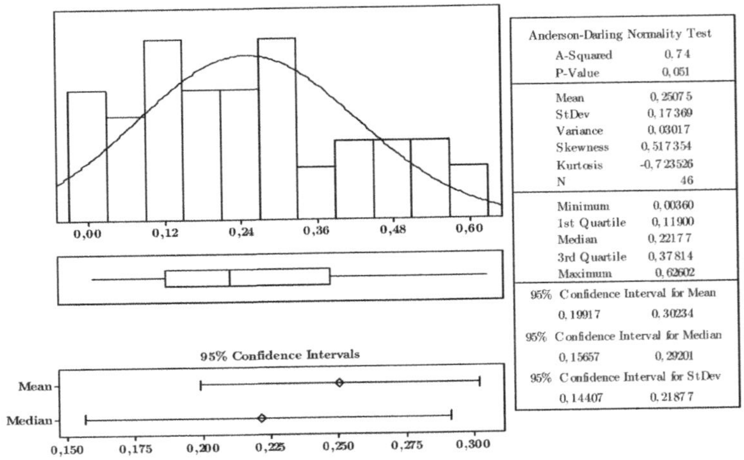

Figure 6.5: Summary statistics of test sample for absolute percentage errors of cost framework #3 via linear regression models

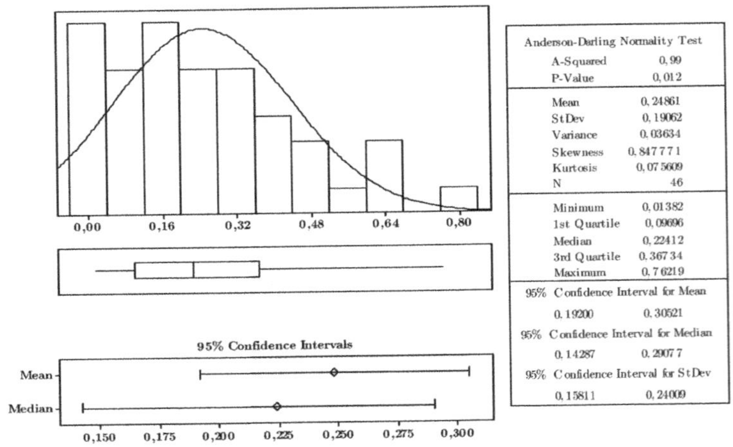

Figure 6.6: Summary statistics of test sample for absolute percentage errors of cost framework #3 via artificial neural network models

When Framework #3 was implemented with artificial neural networks (ANN) as the method of analysis, 3 objects were labeled as outliers and excluded from the test sample. Summary statistics of remaining 46 observations were computed and presented in Figure 6.6. In consistent with Anderson Darling test, it can be argued that normality of the error terms were disturbed; in fact absolute percentage error terms were positively skewed. This suggests considering median along with mean as the measure of central tendency. Mean absolute percentage error of the test sample was calculated as 25% when ANN method in accordance with the flow of information in Framework #3 was considered. Median value, on the other hand, was given as 22%. Dispersion in the sample was computed as 19% which was greater than Framework #2 and equal to Framework #1. Last but not least, the true value of mean of absolute percentage errors shall line between 19% to 31% with confidence level of 95%. This is a wider range in comparison to Framework #1 and #2 when ANN was adopted as method of analysis.

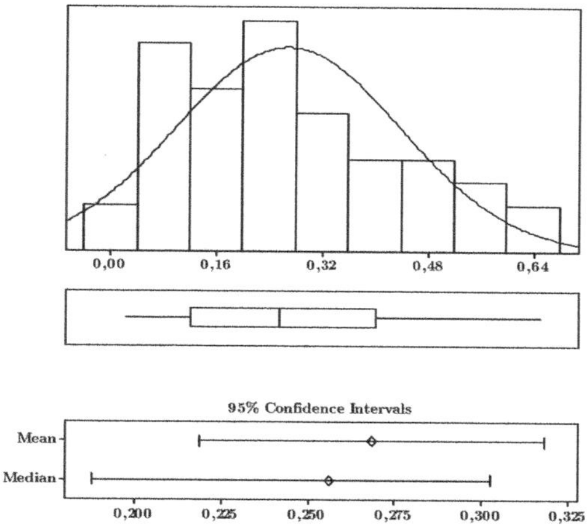

Anderson-Darling Normality Test	
A-Squared	0,70
P-Value	0.062
Mean	0,26866
StDev	0,16952
Variance	0,02874
Skewness	0,526269
Kurtosis	-0,654368
N	47
Minimum	0,02478
1st Quartile	0,12205
Median	0,25648
3rd Quartile	0,40348
Maximum	0,65276
95% Confidence Interval for Mean	
0,21889	0,31843
95% Confidence Interval for Median	
0,18776	0,30281
95% Confidence Interval for StDev	
0,14086	0,21291

Figure 6.7: Summary statistics of test sample for absolute percentage errors of cost framework #4 via linear regression models

Last framework which aimed at cost estimation, Framework #4, adopted *one-step ahead* approach. In Framework #4, an estimator was restricted to employ limited amount of information available HOAI basic evaluation stage. Framework #4 targeted direct estimation of cost of structure (k300+k400). Thereof, no predicted values of cost groups were aggregated to reach cost of structure. When linear regression (LRM) was adopted as the method of analysis, 2 objects were labeled as outliers and excluded from the test sample. Summary statistics of remaining 47 objects were demonstrated in Figure 6.7. Accordingly one could see normality of absolute percentage error terms was not violated. Thereof, estimation of mean value of absolute percentage errors (MAPE) would provide the necessary information regarding central tendency of the true population. MAPE of the test sample when LRM was adopted in accordance with the flow of information in Framework #4 was given in Table 6.1 as 27%. With 95% confidence level, any value of absolute percentage errors in true population was expected to fall between 22% and 32% when Framework #4 was adopted. Lastly, absolute percentage error of 29 projects out of 47 were below 30% in the test sample (Table 6.1).

Framework #4, detected 4 objects in the test sample as outliers when artificial neural network method was considered for the analysis (Table 6.1). Summary statistics of absolute percentage errors (APE) of remaining 45 objects in the test sample were outlined in Figure 6.8. It was evident by Anderson Darling test that normality was violated for APE terms of the test sample. Thereof, *median* of the sample was considered along with *mean* value as the measure of central tendency for APEs. Mean of absolute percentage errors (MAPE) of the test sample was computed as 31%. If the test sample is true representative of population, than statistics suggests MAPE of true population would fall between 23% and 39%, when ANN method in accordance with flow of information in Framework #4 was adopted. This represented relatively wider interval as well as the values were shifted in positive direction when compared to other cost estimation frameworks. The underlying reason was relatively higher amount of dispersion in APE of the test sample. Standard deviation of the test sample was computed as 26%. Another measure of central tendency, median, was calculated as 24%. Last but not least, absolute percentage error of 26 projects out of 45 were below 30% in the test sample (Table 6.1). This is the lowest amount observed among all cost estimation frameworks.

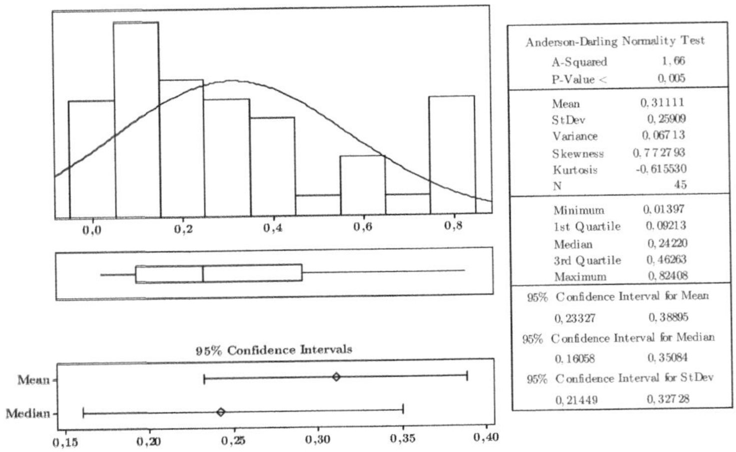

Figure 6.8: Summary statistics of test sample for absolute percentage errors of cost framework #4 via artificial neural network models

The collective box-plots of absolute percentage errors of the test sample with respect to different frameworks and method of analysis were illustrated in Figure 6.9. In conclusion, the lowest *median* value (can be tracked by the horizontal line in the box plots) of the absolute error terms in the test sample was computed when linear regression model in accordance with flow of information in Framework #1 - *multi-way approach* - was adopted. The smallest *mean* value (can be tracked by ∅ symbol in the box plot) of absolute percentage errors was calculated when Framework #2 and linear regression method were adopted at the same time. According to Figure 6.9 and Table 6.1, it can be argued that artificial neural networks did not provide superior results when compared to linear regression methods. Reversely, especially in multi-way approach, linear regression models offered superior prediction accuracy than artificial neural networks.

A formal test of hypothesis, was conducted to reveal whether differences among MAPEs of the test sample is statistically significant at 5% level. This is to investigate if the outcomes of this section can be generalized to overall population or the role of chance can not be ignored. ANalysis Of Variance (ANOVA) was selected as the appropriate statistical procedure to run the formal test of hypothesis. The results of ANOVA concluded that there are *no* differences among the MAPE of construction cost frameworks over test sample at the 5% level of significance. The null hypothesis can not be rejected. The result shall be interpreted in this way: Employing other samples in the population of interest, the chances are greater than 5% that true value of MAPE would fall within the same *interval,* when different methods of analysis in accordance with flow of information in offered frameworks are concerned. This is a problem mostly caused by the size of the test sample. According to statistical theory, as the number of objects in the test sample increases than ability to assess statistical significance of smaller differences also increases.

Last but not least, although statistics do not detect significant differences, for practical implementations 9% improvement can be a very valuable contribution to conceptual cost estimates when linear regression method in accordance with the flow of information in Framework #1 was adopted. Moreover, Framework #1 offers predictions of not only cost of structure but also reference unit quantities and second level cost groups, which can serve as a basis of benchmark at this stage of developments. In conclusion, we argue Framework #1 that adopts multi-way

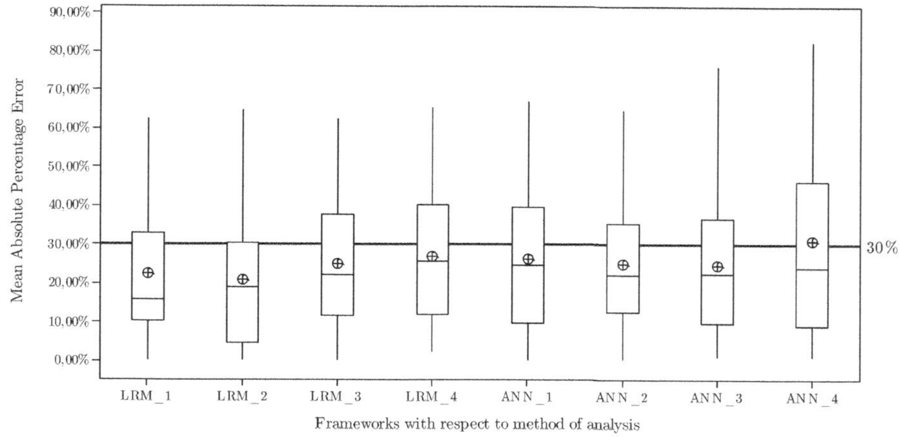

Figure 6.9: Comparison of cost estimation frameworks with respect to method of analysis

approach as the best solution to cost estimation practice at conceptual stage of developments for the population of interest, in this case German building projects.

6.2 Duration Estimation Frameworks

Two frameworks, Framework #5 and #6, were offered to predict construction duration, which was measured in calendar days. Both frameworks employed two methods of analysis that were adopted in this research, namely linear regression (LRM) and artificial neural networks (ANN). This led to reach 4 distinct predictions of construction duration.

Framework #5 adopted *multi-way* approach. In multi way approach, first, estimated values of reference unit quantities were fed back to the system inputs. These cumulative set of input variables were used to model and predict cost of second level cost groups. Following the system inputs were fed back once again with the fitted values of second level cost groups of structure. At last, all collective information were used as inputs to explain the highest possible variance in construction duration. From another perspective, Framework #5 could be imagined as an extension to Framework #1 with additional feedback loop. Framework #6, on the other hand, adopted *one-step ahead* approach in which input variables were restricted according to the assumption of this study. Recall, the inputs were limited to a small set of information, macro project determinants, at HOAI basic evaluation stage.

The identical test sample, which was also used to validate prediction performance of cost estimation frameworks, consisted of 49 objects was employed. The performance measurements were considered as root mean square error (RMSE) and mean absolute percentage error (MAPE) like the rest of the study. Collective results of construction duration frameworks with respect to method of analysis were outlined in Table 6.2.

According to Table 6.2, Framework #5 detected 5 observations as outliers when linear regression (LRM) was considered as the method of analysis. Individual investigation of absolute

Table 6.2: Framework comparison of the duration models

Method of analysis	Framework #	Performance measurements			
		RMSE	MAPE	n*	N*
LRM	5	214.28	30.30%	5	23
	6	249.59	33.64%	4	19
ANNs	5	215.78	29.32%	5	21
	6	208.64	28.14%	5	24

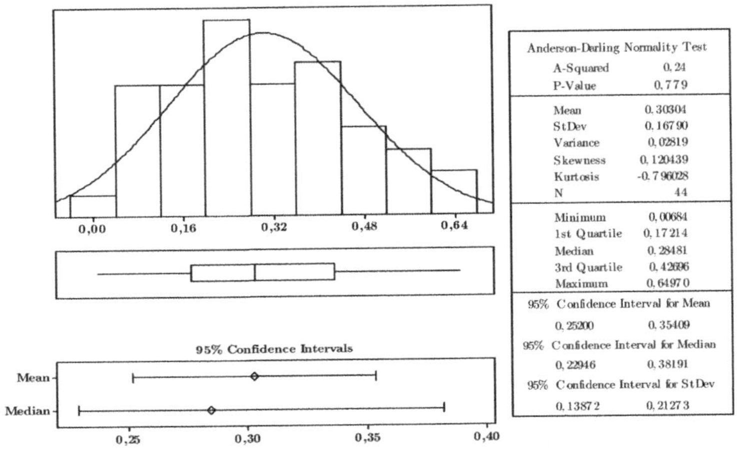

Figure 6.10: Summary statistics of test sample for absolute percentage errors of construction duration framework #5 via linear regression models

percentage errors (APE) in the test sample was performed. Summary statistics of this investigation were illustrated in Figure 6.10. Anderson Darling test suggested that normality of error terms could be assumed. Thereof, mean value of absolute percentage error (MAPE) indicated a fair measure for central tendency of the test sample. Remaining 44 objects supplied MAPE of 30%. Measure of dispersion, standard deviation, was calculated approximately 17%. One could estimate the interval in which the values of true population fall. The results suggested with 95% confidence that absolute percentage error of duration estimation would be between 25% and 35%, when linear regression along with flow of information offered by Framework #5 was considered. Lastly, APE of 23 out of 44 objects were computed below 30% (Table 6.2).

When artificial neural networks (ANN) was considered as the method of analysis, Framework #5 detected equal number of observation as outliers as LRM. The summary statistics of individual absolute error terms (APE) of the test sample was given in Figure 6.11. It was evident by Anderson Darling statistic that probability distribution of APEs was normal. Mean of absolute percentage errors (MAPE) was computed as 29%. This indicated very slight improvement in comparison to LRM. On the other hand dispersion of absolute percentage errors of the test sample was 19%. This suggests that values of APE lied in a wider range than LRM, when ANN models were considered for prediction. 95% confidence interval for mean absolute percentage error was estimated between 23.5% and 35%. This indicated slightly a wider interval in comparison to LRM, influenced by higher value of standard deviation. Lastly, absolute percentage error of 19 objects in the test sample was computed below 30% (Table 6.2).

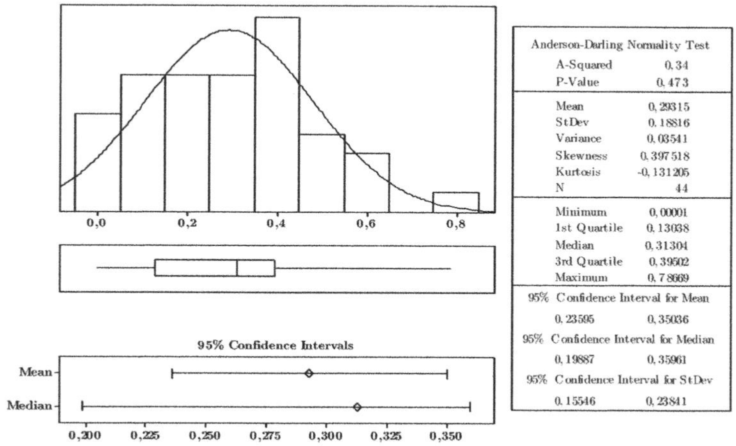

Figure 6.11: Summary statistics of test sample for absolute percentage errors of construction duration framework #5 via artificial neural network models

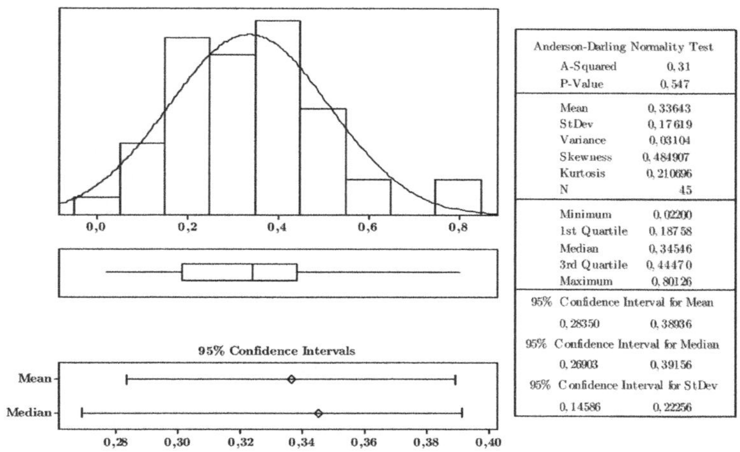

Figure 6.12: Summary statistics of test sample for absolute percentage errors of construction duration framework #6 via linear regression models

The second framework, of this section, and last, of this research, Framework #6 adopted *one-step* ahead approach. As mentioned previously, it was assumed that the input variables were restricted to those which were available to an architect at HOAI basic evaluation stage. This was initially offered for comparative aims. Linear regression analysis detected 4 observations as outliers in the test sample. Individual investigation of absolute percentage errors (APE) was presented in Figure 6.12. Accordingly, one can see in Figure 6.12 that normality was not violated for the APE terms of the test sample. Thereof, *mean,* as the measure of central tendency, was considered. MAPE of the test sample was computed as 34%. In comparison to multi-way approach (Framework #5), the result was 4% higher, which indicated poorer prediction ability. Considering computed mean as the estimate of the population, one can compute that the true value of the population positioned between 28% and 39%, at confidence level of 95%. Lastly, absolute percentage error of 21 objects were computed below error rate of 30%.

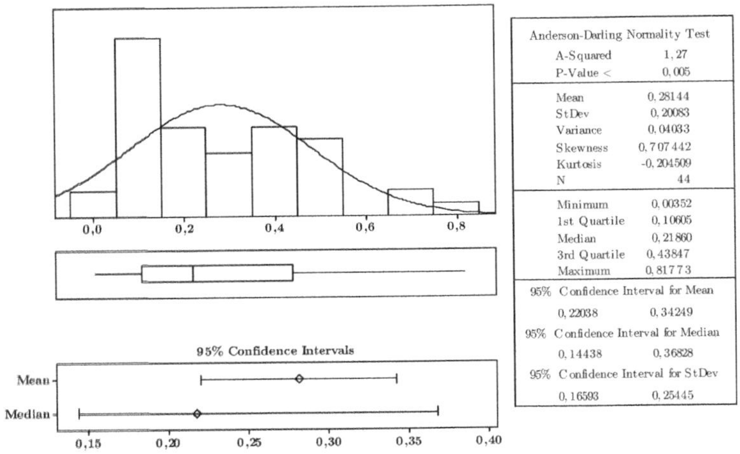

Figure 6.13: Summary statistics of test sample for absolute percentage errors of construction duration framework #6 via artificial neural network models

Framework #6 detected 5 objects as outliers when artificial neural network (ANN) was considered as the method of analysis. These objects were excluded from the test sample. Absolute percentage errors of remaining 44 objects were analyzed and the result were illustrated in Figure 6.13. Anderson Darling normality test suggested that APEs of the test sample did not follow normal distribution. Thereof, median value was considered along with mean, as the measures of central tendency. Mean of absolute percentage error (MAPE) was computed as 28%, whereas the median was determined as 22%. It can be tracked in Figure 6.13, couple of projects were located at the right tale of the distribution, not outliers, but violate normality. It was evident these projects had influence on the mean value, to an extent. Standard deviation, as measure of dispersion, was computed as 20%. Relatively larger dispersion of APE values of the test sample influenced the range of prediction intervals. Considering computed mean as the true estimation of the population, one could expect absolute percentage error of duration estimation would fall between 22% and 34%, when ANN method in accordance with flow of information in Framework #6 was considered. Last but not least, analysis pointed 24 objects out of 44 reaching to APE below 30%.

Analysis of variance (ANOVA) was implemented to test the null hypothesis of no differences among the means of absolute percentage errors over the test sample were observed. The results by the formal test of hypothesis concluded that the null hypothesis could not be rejected at 5% significance level. Therefore it was concluded that the difference among groups means was not significant.

Box-plots of absolute percentage errors of the test sample with respect to construction duration frameworks and method of analysis was given in Figure 6.14. It can also be visually detected in Figure 6.14 that different frameworks offered very close values of mean absolute percentage errors. Thereof, different considerations shall be taken into account. Figure 6.14 showed the Framework #6 offered the lowest value of median, when ANN was adopted as the method of analysis. On the other hand, it was evident with the findings that Framework #5 supplied the smallest dispersion in absolute percentage errors of the test sample when LRM was considered. However, it could be argued Framework #5 offered a critical advantage when

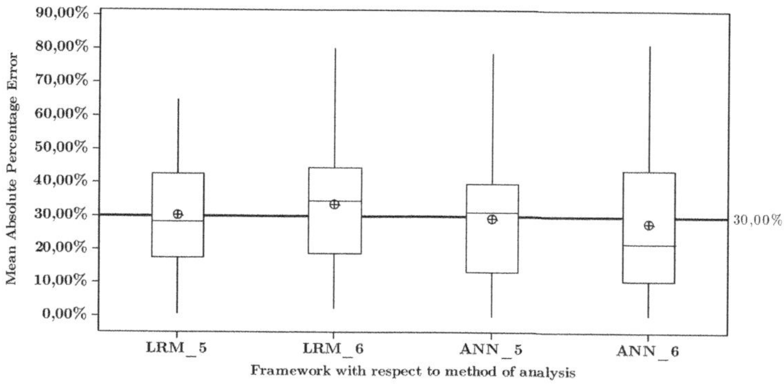

Figure 6.14: Comparison of construction duration frameworks with respect to method of analysis

compared to its counter part, Framework #6. Multi-way approach, which was adopted by Framework #5, supplied estimation of not only construction duration but also reference unit quantities and cost groups of structure. Moreover, expected errors would be more consistent due smaller standard deviation. In conclusion, Framework #5 with linear regression method was nominated for those who aimed at predicting construction duration.

7 Implementation

Chapter 7 aimed to illustrate practical implementation of the results. Depending on conducted analysis (Chapter 5) and comparisons (Chapter 6); the findings of the study suggested cost estimation Framework #1 and construction duration Framework #5 offered higher accuracy in prediction and more consistency in error. Thereof, these frameworks were utilized to exemplify implementation of the outcomes.

First, a building project that had been included in the test sample was randomly selected. This indicated that this project information had not been included in modeling or, so called, training process, previously. Therefore, aim of assessing ability of pattern recognition could be measured objectively, and presented in this demonstration. Selection process determined the test object with BKI ID no. 1300-082, which has been a *private office* project. Following, point estimations (expected average values) of reference unit quantities, second level cost groups and construction duration, according to the flow of information in Frameworks #1 and #5, were calculated. Individual absolute percentage errors of each computation were reported. prediction intervals of the these point estimations were also supplied. This enabled to quantify the amount of insecurity inherent in the prediction.

7.1 Story of 1300-082

Please consider yourselves traveling back in time to somewhere in 1995. There, you were an architect who were recently approached by a client for an office design and asked to determine strategic targets of the upcoming project. It was known of very few information regarding the building to be constructed. The design of the building which was to be executed by you had no details or what so ever. The project is through the phase of establishing the basis.

Through establishing the basis of a project, few information that the architect could be sure of with some confidence (Table 7.1) was available. Type of client which has administrated the project was "private". Type of facility to be built, office, was provided by the client at the initial approach. Commence date of construction was planned to be sometime in 1997. Subsequent to a short investigation of the purchased plot of development, topography of the plot and site access to the construction site was identified as "flat" and "free", respectively. The general pattern of the ground conditions in the plot region, which was provided by local authorities, was "moderate". This corresponded to soil class of 3 to 5 according to DIN 18300. Observed declining trend of the construction market led to define market conditions as "stagnation".BKI location index was not open to interpretation and defined clearly with respect to location as 1.0008. In accordance with the client's requirements as well as regulative restrictions by municipal institutions, gross external floor area and gross building volume was approximately computed as 8,000 $m^2$and 32,000 m^3, respectively. The very same regulations restricted the development to have 4 storeys above ground and 1 storey below ground, which aggregated to 5 storeys in total for this particular plot. Using these quantities average storey height, the ratio of gross building volume to gross external floor area, average floor size, the ratio of gross external floor area to number of storeys were calculated as 4 m and 1,600 m^2, respectively. Last but not least, plot area, 7,906 m^2, and allowed surface to develop the building on, 1,921 m^2, were available to the architect.

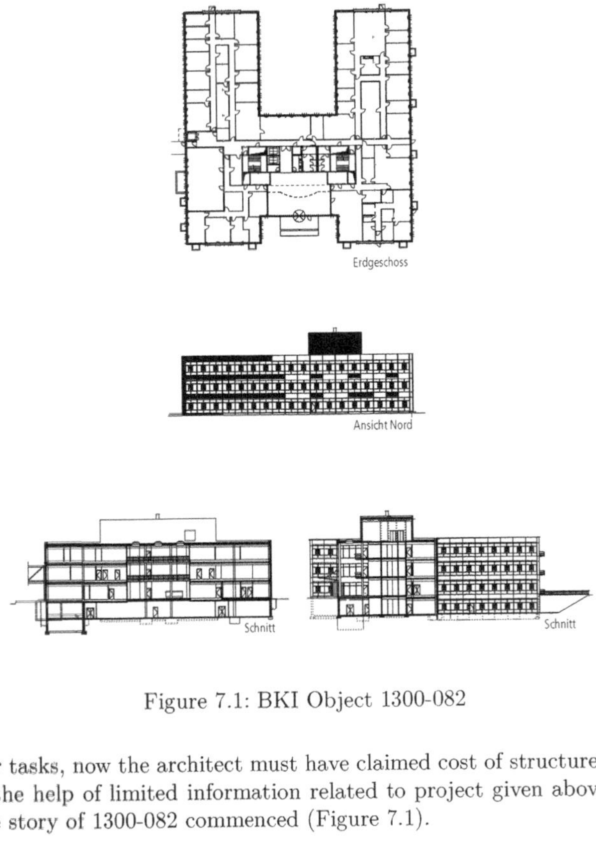

Figure 7.1: BKI Object 1300-082

Among other tasks, now the architect must have claimed cost of structure and construction duration, with the help of limited information related to project given above. This was how, *we assumed*, the story of 1300-082 commenced (Figure 7.1).

7.2 Point Estimates

As indicated previously, Framework #1 and #5 were selected to demonstrate the practical usage of resulted models. Given the fact that there were no significant differences observed between test sample means for absolute percentage error of different method of analysis; both multiple linear regression and artificial neural network models were employed to calculate the values of interest.

Table 7.1: Model inputs, that were available to the architect, of 1300-082 at the basic evaluation stage of HOAI.

Model inputs	BKI Object ID
	1300-082
FT	office
CSD	1997
CT	private
TOP	"flat" - (1)
SA	"free" - (1)
GC	"moderate" - (2)
MC	"stagnation" - (1)
LI	1.0008
GEFA	8,000 m^2
GBV	32,800 m^3
NoS	5
NoS_{ag}	4
NoS_{bg}	1
ASH	4.06 m
AFS	1,615.20 m^2
PA	7,906.00 m^2
DPA	1,921.00 m^2

Recall Figure 1.4 and 1.5. The flow of information in employed frameworks, thereof through practical implementation, was as follows:

1. Prediction of reference unit quantities
2. Feedback system inputs with predicted (fitted) values of reference unit quantities
3. Prediction of second level cost of construction works
4. Feedback system inputs with predicted (fitted) values of second level cost of construction works
5. Prediction of construction duration

7.2.1 Reference unit quantities

According to DIN 277-3 (2005) reference unit quantities refers to the structural elements of a building: Namely, excavation volume (m310), foundation area (m320), external walls area (m330), internal walls area (m340), floors and ceilings area (m350), and roof area (m360). DIN 276-1 (2008) identifies these structural elements to be used as unit measures through identification of second level cost of construction works, structure.

Relying on this information, it was previously hypothesized that estimation of these quantities with some confidence may cause an increase in prediction accuracy of estimations of second levels cost of structure, to an extent. One could see in Table 6.1 that improvement in measure of prediction accuracies was gained, to an extent. However, formal test of hypothesis concluded that these differences were not significant at 5% level.

In Chapter 5, linear equations that best predicts the reference unit quantities of office buildings were determined as follows (Table 5.2)

$$m310^{0.209378} = 2.10957 + (0.000272976 \times GEFA) - (3.50077E-05 \times GBV) + (0.0860543 \times NoS_{ag}) + (0.000796656 \times AFS) + (1.2872 \times LI) + (0.245398 \times TOP)$$

$$m320^{0.847661} = 17.7126 + (0.199476 \times DPA) + (0.060917 \times AFS) + (0.00960183 \times GEFA) + (7.58206 \times ASH)$$

$$m330^{0.599167} = 7.41863 + (0.0113117 \times GEFA) + (5.52124 \times NoS) + (0.000155539 \times PA) - (0.00183323 \times DPA) + (0.00480411 \times AFS) + (4.13878 \times ASH) + (1.7675 \times TOP)$$

$$m340^{0.5} = 8.44356 + (0.00828796 \times GEFA) - (0.000671205 \times GBV) + (3.10747 \times NoS_{ag}) + (2.7164 \times NoS_{bg}) + (0.000126272 \times PA) + (0.00407928 \times AFS)$$

$$m350^{0.892337} = 117.21 + (0.345654 \times GEFA) + (59,5652 \times NoS_{bg}) - (0.273902 \times DPA) - (0.0843562 \times AFS) - (15.8063 \times TOP)$$

$$m360^{0.716128} = 27.5193 + (0.00556912 \times GEFA) + (0.0395838 \times DPA) + (5.94156 \times ASH) + (0.0507919 \times AFS)$$

To obtain a value of interest one must insert relevant model inputs, which was listed in Table 7.1, to an equation.

External walls area (m330) was calculated to exemplify a solution. According to the model developed, gross external floor area (GEFA), number of storeys (NoS), plot area (PA), developed plot area (DPA), average floor size (AFS), average storey height (ASH) and topography class (TOP) were significant factors at 5% level which explained the highest possible portion of variation in external walls area. When numerical values of these variables were read from Table 7.1 and inserted the equation:

$$m330^{0.599167} = 7.41863 + (0.0113117 \times 8000) + (5.52124 \times 5) + (0.000155539 \times 7906) - (0.00183323 \times 1921) + (0.00480411 \times 1600) + (4.13878 \times 4) + (1.7675 \times 1)$$

$$m330^{0.599167} = 150.416721$$

$$m330 = \sqrt[0.599167]{150.41672} = 150.41672^{1/0.599167}$$

$$m330_{fit} = 4303.99 \cong 4304\text{m}^2$$

Another implementation of the result could be to generate reference unit quantity indicators, which was widely used in practice by German architects, instead of using raw values. An indicator was calculated simply as the ratio of a reference unit quantity to gross external floor area:

$$Indicator = \frac{Raw\,quantity}{Gross\,external\,floor\,area} \tag{7.1}$$

Accordingly, predicted value of external wall area indicator for linear regression models was calculated as

$$m330_{LRM,fit,ind.} \quad = \quad {}^{4304}/_{8076} = 0.533$$

Explained in detail in Chapter 3, Artificial Neural Networks (ANN) could be assumed as computational devices that could be simulated using software applications. That was why ANNs were referred to be *black box* models.

Hence the aim of this research did not incorporate with developing Neural Network software but rather testing their capabilities of pattern recognition for the problem of interest, in this case prediction of project determinants; a commercial software was utilized.

Table 7.2: Predicted values of reference unit quantities for 1300-082

Response	Actual		Predicted (fits)				*Abs. Error (%)*	
			LRM		ANN			
	Raw	Ind.	Raw	Ind.	Raw	Ind.	*LRM*	*ANN*
m310 (m^3)	19,394	2.401	6,724	0.833	18,489	2.289	*65.33*	*4.67*
m320 (m^2)	1,921	0.238	1,923	0.238	1,979	0.245	*0.09*	*3.03*
m330 (m^2)	4,718	0.584	4,304	0.533	4,508	0.558	*8.76*	*4.45*
m340 (m^2)	7,802	0.966	5,790	0.717	5,364	0.664	*25.79*	*31.25*
m350 (m^2)	6,155	0.762	5,824	0.721	5,542	0.689	*5.38*	*9.96*
m360 (m^2)	1,996	0.247	2,290	0.284	2,537	0.314	*14.72*	*27.12*

Identical factors that was found to make significant contribution to linear regression prediction were used as input nodes of ANN model. In order to enable a fair comparison to linear regression models, identical training and test samples were utilized as well. Multi layer feed-forward (MLF) net, with 4 neurons in its hidden layer, provided the smallest error on the test sample for the prediction of external walls area (Table 5.4). Relative variable impacts of input nodes for external wall area prediction was given in Figure 7.2. ANN model predicts external wall area of 1300-082 as 4,508 m^2. Using equation 7.1 the indicator of this value was given as:

$$m330_{ANN,fit,ind.} \quad = \quad {}^{4508}/_{8076} = 0.558$$

Actual external wall area of 1300-082, that has been incurred in the construction site, was observed as 4,718 m^2. Actual value of external wall area indicator was:

$$m330_{act,ind.} \quad = \quad {}^{4718}/_{8076} = 0.584$$

Absolute percentage error (APE) of estimation was computed as:

$$APE = \frac{|\,Actual - Fit\,|}{Actual} \times 100 \tag{7.2}$$

For external wall area, absolute error of linear regression and ANN prediction were given below, respectively

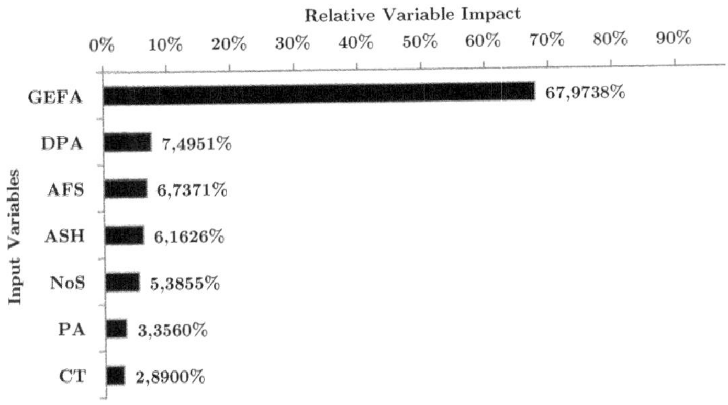

Figure 7.2: Relative impacts of input nodes for external wall area prediction via MLF net

$$APE_{m330,LRM} = (|4718-4303|/4718) \times 100 = 8.76\%$$

$$APE_{m330,ANN} = (|4718-4508|/4718) \times 100 = 4.45\%$$

Repeating the same exercises for each of the remaining reference unit quantities, one could estimate the values of interest that was given in Table 7.2.

7.2.2 Cost of structure

Subsequent to prediction of values for expected average reference unit quantities, the architect had more information regarding 1300-082 with *some confidence*. According to the hypothesis, the architect could use these additional information along with the initial inputs, to estimate 1300-082's cost of structure more precisely.

Another advantage of employing offered multi-way forecast framework would be emphasized at that point. As mentioned previously, cost of structure was equal to the sum of its 17 sub-groups of cost, where 8 of them, namely cost of excavation (k310), foundation (k320), external walls (k330), internal walls (k340), floors and ceilings (k350), roof (k360), structural fitments (k370), other construction related activities (k390), formed cost structure - construction works (k300) and remaining 9, namely cost of sewerage water and gas systems (k410), heat supply systems (k420), air treatment systems (k430), power installations (k440), telecommunications and other communication systems (k450), transport systems (k460), function-related equipment and fitments (k470), building automation (k480), other service related work (k490), formed cost of structure - services (k400). Although the project presented no details of design, multi-way forecast framework offered *expected average values* of these cost groups with some confidence.

Linear models that explained the highest possible portion of variance in cost of structure were given in Table 5.6 and 5.7, respectively. One could construct the equation for the value of interest for *private office developments* as follows:

$$k310^{0.109225} = 2.45633 + (2.90669E-05 \times m310_{fit}) + (0.263793 \times NoS_{bg}) + (0.129809 \times SA) + (0.0708819 \times MC) + (0.00124581 \times DPA) + (3.0804E-05 \times GBV) - (0.00144288 \times m320_{fit})$$

$$k320^{0.370276} = 56.6972 + (0.016338 \times m320_{fit}) + (0.00131328 \times GBV) + (2.31535 \times MC) + (0.000223414 \times PA)$$

$$k330^{0.0736077} = 2.37302 + (0.000233058 \times m330_{fit}) - (7.59021E-05 \times m340_{fit}) - (0.0162469 \times NoS_{ag}) + (0.000283306 \times DPA) - (0.000506556 \times m360_{fit}) + (0.000347912 \times AFS) + (0.0366708 \times NoS)$$

$$k340^{0.325125} = 27.8763 + (0.0244354 m330_{fit}) + (0.0183367 \times GEFA) + (2.15322 \times MC) - (0.0168308 \times m340_{fit}) - (0.00225432 \times GBV)$$

$$k350^{0.38558} = 31.1209 + (0.0671908 \times m350_{fit}) - (0.0355633 \times m340_{fit}) + (0.00567296 \times m310_{fit}) + 0.027788 \times m360_{fit}) + (0.000643071 \times PA) - (0.0262668 \times m330_{fit}) + (14.717 \times NoS) + (2.78606 \times SA)$$

$$k360^{0.114241} = 3.277 + (0.00133087 \times m360_{fit}) + (0.000295679 \times AFS) + (0.0655223 \times MC) + (2.85474E-06 \times PA) + (0.00003 \times m310_{fit}) + (0.0527771 \times NoS) + (0.0031414 \times DPA) - (0.00459676 \times m320_{fit})$$

$$k370^{0.157933} = 4.62793 - (0.200909 \times NoS_{ag} + (0.000233043 \times m350_{fit}) + (5.48127 \times E - 05 \times GBV) + (0.321014 \times NoS_{bg})$$

$$k390^{0.153078} = 3.8787 + (0.000810971 \times m330_{fit}) - (0.000285152 \times m340_{fit}) + (0.195333 \times NoS_{ag}) + (0.000379667 \times AFS) + (0.183581 \times NoS_{bg})$$

$$k410^{0.16313} = 5.28366 - (0.000154199 \times m310_{fit}) + (0.000814858 \times GEFA) - (0.00053767 \times m330_{fit}) + (0.250692 \times NoS_{bg})$$

$$k420^{0.219709} = 7.4826 + (0.000718509 \times GEFA) + (0.168898 \times NoS_{ag}) + (0.000604654 \times DPA + (2.85717 \times LI)$$

$$k430^{0.0970641} = 2.16166 - (2,82536E-05 \times GBV) + (0.000313817 \times AFS) - (0.000253741 \times m340_{fit}) + (0.00075723 \times m330_{fit})$$

$$k440^{0.11659} = 2.87219 - (8.55834E-05 \times GBV) - (0.000193788 \times AFS) + (0.000677079 \times GEFA) - (0.000557373 \times m340_{fit}) + (5.22724E-06 \times PA) - (0.06414 \times TOP) + (0.000877612 \times m330_{fit}) + (0.0829667 \times ASH) - (0.000202217 m350_{fit}) + (0.0529722 \times MC)$$

$$k450^{0.0480147} = 1.53653 - (0.00147849 \times m350_{fit}) - (0.000104095 \times DPA) + (0.0123983 \times NoS) + (0.000155578 \times GEFA)$$

$$k460^{0.359114} = 41.3989 + (0.0210634 \times m330_{fit}) - (0.00908751 \times AFS) - (0.00660191 \times m340_{fit})$$

$$ln(k470) = 8.70902 + (0.000952183 \times GEFA) - (0.255767 \times NoS_{ag}) - (0.000583192 \times m340_{fit})$$

$$k480^{0.5} = 664.657 + (0.0113522 \times PA) + (0.0464505 \times GEFA) - (785.332 \times LI) + (100.092 \times MC)$$

$$ln(k490) = 7.22284 + (0.000373859 \times GEFA)$$

We went on demonstrating our example with cost of external walls. Linear model suggested not only fitted value of external wall area but also fitted values of internal wall area, roof area, along with number of stories above ground, developed plot area, average floor size and number of stories as the significant predictors for cost of external walls. Note that three of these predictors were resulted from the first stage of our analysis.

To calculate the fit for cost of external walls, significant predictor values would be inserted to the linear equation as follows:

$$k330^{0.0736077} = 2.37302+(0.000233058\times4304)-(7.59021E-05\times5790)-(0.0162469\times4)+(0.000283306\times1921)-(0.000506556\times2290)+(0.000347912\times1600)+(0.0366708\times5)$$

$$k330^{0.0736077} = 3.001381$$

$$k330 = \sqrt[0.0736077]{3.001381} = 3.001381^{1/0.0736077}$$

$$k330_{fit} = 3,052,560€$$

And its cost indicator for linear model was computed as:

$$k330_{LRM,fit,ind.} = 3{,}052{,}560€/8{,}076m^2 = 377.98€/m^2$$

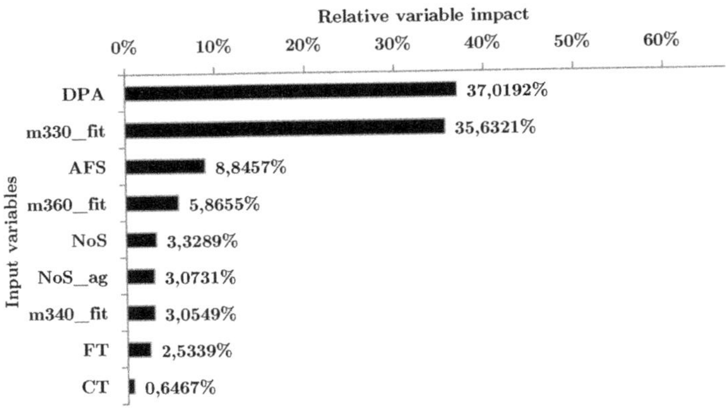

Figure 7.3: Relative impacts of input nodes for cost of external wall area prediction via MLF net

Employing the significant inputs of linear model, ANN model was trained for external walls area as in reference unit quantity estimation. Multi layer feedforward net with 7 nodes in its hidden layer minimized the root mean square error of the test sample (Table 5.9). Relative variable impacts of artificial neural network model inputs were shown in Figure 7.3. Best ANN model predicted cost of external wall of 1300-082 as 2,786,318 €. Following, cost indicator of ANN model was calculated as:

$$k330_{ANN,fit,ind.} \quad = \quad {}^{2{,}786{,}318€}/_{8{,}076\mathrm{m}^2} = 345.01€/\mathrm{m}^2$$

Subsequent to practical completion of the construction, the architect was responsible to conduct statement of final costs (*kostenfeststellung*) where he identified the actual costs incurred in the construction site. Actual value for the cost of external walls (k330) was identified as 2,439,705 €. Its ratio to gross external floor area, the indicator, was calculated as 302.09 €/m² by using equation 7.1. Later, employing equation 7.2, absolute percentage error of the architect's initial estimation could be calculated as:

$$APE_{k330,LRM} \quad = \quad ({}^{|377.98-302.09|}/_{302.09}) \times 100 = 25.12\%$$

$$APE_{k330,ANN} \quad = \quad ({}^{|345.01-302.09|}/_{302.09}) \times 100 = 14.21\%$$

Repeating the same steps of calculation for remaining cost groups of structure, it was possible for the architect to predict expected values of all related cost groups. Please recall, it was previously assumed that the architect was competent enough to assess the cost groups which did not require construction work and therefore valued as zero. In the case of 1300-082, cost of other service related work (k490) fitted that assumption.

Collective results for comparison between actual values and predictions (fits) calculated by linear regression and artificial neural networks models of cost groups of structure for 1300-082 was presented in Table 7.4.

Cost of structure - construction works (*k300**), which equaled to the aggregate of its sub cost groups (k310+...+k390), was computed as 9,472,230 € and 8,243,060 € by linear regression and artificial neural networks, respectively. Whereas the actual cost was observed as 7,559,098 €. Utilizing equation 7.1 and 7.2 one could find their cost indicators and associated absolute percentage errors. Accordingly, linear regression models results in significantly higher error, 25.31%, when compared to artificial neural network models whose aggregated error was 9.05% (Table 7.4). It shall also be noted that both method of analysis overestimated.

Cost of structure - services (*k400**), on the other hand, showed a similar pattern in terms of comparison between method of analysis: Artificial neural nets provided superior accuracy when compared to linear models. Yet, this time neither linear regression nor artificial neural network models overestimated; instead, they underestimated aggregated cost by 32.41% and 18.16%, respectively (Table 7.4).

Finally, aggregated cost of structure (*k300*+k400**) reached to a very high level of prediction accuracy for both method of analysis. In the case of 1300-082, it was evident that error of overestimation in cost of construction works balanced the error of underestimation in cost of services. This led to prediction accuracies which could be regarded as *more than ideal* at the conceptual phase of design, reaching the error of 2.97% and 1.48% by linear regression and artificial neural nets, respectively (Table 7.4).

If predicted values and associated errors in Table 7.4 would be investigated in detail, than it could be seen that 10 out of 15 cost groups reached the prediction error less than 30% with linear regression method. Whereas the number shrunk to 6 out of 15 when artificial neural network method was considered. This demonstrated that the results of linear regression method provided more ideal distribution of individual errors of second level cost groups, such as cost of external walls, in comparison to artificial neural network method. By other words, it

Table 7.4: Predicted values of cost groups of structure for 1300-082

Response	Actual		Predicted (fits)				Abs. Error (%)	
			LRM		ANN			
	Raw (€)	Ind. (€/m²)	Raw (€)	Ind. (€/m²)	Raw (€)	Ind. (€/m²)	LRM	ANN
k310	237,514	29.41	178,250	22.07	146,206	18.10	*24.95*	*38.44*
k320	468,382	58.00	569,881	70.56	788,582	97.65	*21.67*	*68.36*
k330	2,439,705	302.09	3,052,560	377.98	2,786,318	345.01	*25.12*	*14.21*
k340	1,613,306	199.77	2,002,637	247.97	1,124,520	139.24	*24.13*	*30.30*
k350	1,836,746	227.43	2,362,352	292.52	1,718,824	212.83	*28.62*	*6.42*
k360	503,369	62.33	578,663	71.65	767,314	95.01	*14.96*	*52.44*
k370	204,611	25.34	292,890	36.27	234,465	29.03	*43.14*	*14.59*
k390	255,464	31.63	434,998	53.86	676,831	83.81	*70.28*	*164.94*
k410	409,557	50.71	600,936	74.41	366,120	45.33	*46.73*	*10.61*
k420	452,667	56.05	514,458	63.70	589,209	72.96	*13.65*	*30.16*
k430	1,106,102	136.96	442,567	54.80	622,394	77.07	*59.99*	*43.73*
k440	1,216,830	150.67	922,381	114.21	1,039,558	128.72	*24.20*	*14.57*
k450	1,062,591	131.57	193,199	23.92	464,538	57.52	*81.82*	*56.28*
k460	219,244	27.15	193,516	23.96	183,023	22.66	*11.74*	*16.52*
k470	187,492	23.22	162,642	20.14	426,206	52.77	*13.25*	*127.32*
k480	118,928	14.73	196.846	24.37	215,599	26.70	*65.52*	*81.29*
k490	-	-	-	-	-	-	-	-
*k300**	*7,559,098*	*936.00*	*9,472,230*	*1,172.89*	*8,243,060*	*1,020.69*	*25.31*	*9.05*
*k400**	*4,773,412*	*591.06*	*3,226,545*	*399.52*	*3,906,648*	*483.74*	*32.41*	*18.16*
k300+k400**	*12,332,510*	*1,527.06*	*12,698,775*	*1,572.41*	*12,149,708*	*1,504.42*	*2.97*	*1.48*

might be that the *cosmic luck* was in the lead act when lower prediction error was reached for aggregated costs of construction works and services with artificial neural networks in the case of 1300-082.

Table 7.5: Analysis of variance for cost of sewerage, water and gas systems (k410) linear model

Source	$d.f$	Seq SS	Adj SS	Adj MS	F	p(F)
Regression	9	495.005	495.005	55.0006	199.520	$< 5\%$
CT	1	46.174	3.220	3.2201	11.681	$< 5\%$
FT	4	24.224	10.398	2.5995	9.430	$< 5\%$
$m310_{fit}$	1	290.501	7.873	7.8726	28.558	$< 5\%$
GEFA	1	125.189	52.993	52.9931	192.237	$< 5\%$
$m330_{fit}$	1	3.244	4.984	4.9838	18.079	$< 5\%$
NoS_{bg}	1	5.674	5.674	5.6742	20.584	$< 5\%$

Another important point was prediction in cost of construction works, for both methods, reached smaller prediction error when compared to cost of services in the case of 1300-082. This did not surprise us hence predicted reference unit quantities which later fed back system inputs in accordance with the principle of multi-way forecast approach, was more relevant to construction works rather than services. Although some of reference unit quantities were also employed as independent variables in cost of services models, the largest portion of variance in these cost groups was always explained by initial inputs, such as gross external floor area, that were listed in Table 7.1. Analysis of variance for cost of sewerage, water and gas systems (k410) linear model was given to exemplify the case in Table 7.5. It was evident with F statistics that gross external floor area explained significantly a higher portion of variance in this cost group (Table 7.5). Besides, the figure showed relevance with the current practice in Germany since reference unit quantity for second level cost of services was identified as gross external floor area (recall Table 1.1). More important, this result suggested *revealing more information* approach worked better in the case of 1300-082, due to smaller error observed in cost of construction works than cost of services.

7.2.3 Construction duration

In consistent with the flow of information in selected frameworks, last task was to perform estimation of construction duration. According to the multi-way forecast approach, the model inputs were fed back by the predicted values of second level cost groups of structure. Thereof, more information was revealed to the architect to perform his estimation.

Linear model that explained the most substantial part of variance in construction duration for private office developments was given in Table 5.20 as:

$$CD^{0.249539} = 16.5035 - (0.00624879 \times CSD) + (0.241099 \times NoS_{bg}) + (1.45705E-06 \times k370_{fit}) + (2.37419E-06 \times k460_{fit}) + (2.04239E-005 \times GBV) - (3.28438E-06 \times k390_{fit}) + (0.109459 \times NoS_{ag})$$

Reading the values of model inputs from Table 7.1 and 7.4 one would construct following equation:

$$CD^{0.249539} = 16.5035 - (0.00624879 \times 1997) + (0.241099 \times 1) + (1.45705E - 06 \times 292,890) + (2.37419E - 06 \times 193,516) + (2.04239E - 05 \times 32,000) - (3.28438E - 06 \times 434,998) + (0.109459 \times 4)$$

$$CD^{0.249539} = 4.8312$$

$$CD = \sqrt[0.249539]{4.8312} = 4.8312^{1/0.249539}$$

$$CD = 551 \text{days}$$

Using the same inputs and training data set, an artificial neural network model was developed. Generalized regression neural net (GRN Net) provided the lowest root mean square error in the test sample and therefore selected to predict the values of unseen data. Relative variable impacts of artificial neural network model inputs were shown in Figure 7.4. Construction duration of 1300-082 was predicted as 627 calendar days by GRN Net.

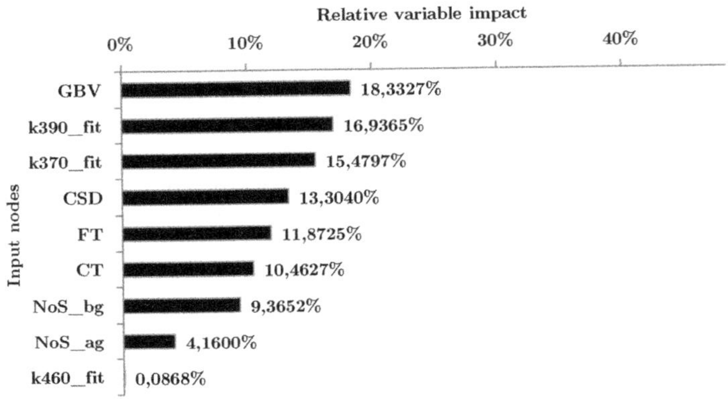

Figure 7.4: Relative impacts of input nodes for construction duration prediction via GRN net

1300-082's actual construction duration was observed as 638 calendar days by the end of construction period. Using the equation 7.2, one could calculate absolute percentage errors of linear regression model as

$$APE_{CD,LRM} = (^{|551-638|}/_{638}) \times 100 = 13.64\%$$

and GRN net as

$$APE_{CD,ANN} = (^{|627-638|}/_{638}) \times 100 = 1.79\%$$

In the case of 1300-082, it was evident that multi-way ANN approach provided significantly better predictive accuracy for the point estimate of construction duration. It was also noted that the multi-way models employed three cost groups that was revealed in the second phase of analysis along with macro determinants of a project, such as gross building volume and number of stories above ground.

It would be underlined, as mentioned before, point estimates of 1300-082 could be regarded as *more than ideal* for a prediction that was performed through establishing the basis of a project. It was evident with the findings presented in Table 6.1 and 6.2, higher error rates are very likely for cost of structure and construction duration when compared to randomly selected case presented as an example of implementation.

7.3 Confidence Intervals

Implementation of proposed models and frameworks for the prediction of project determinants demonstrated that they offer practically applicable results and may form a common basis for estimation of project determinants through establishing the basis of a project belonging to the population of interest, in this case building projects in Germany.

Table 7.6: 95% confidence intervals for expected values of 1300-082

Response	Actual	Expected	95% Confidence Intervals		*Abs. Error (%)*		
			Lower	Upper	*Exp.*	*Low.*	*Up.*
m310 (m^3)	19,394	6,724	5,645	7,960	*65.3*	*70.9*	*59.0*
m320 (m^2)	1,921	1,923	1,874	1,971	*0.1*	*2.4*	*2.6*
m330 (m^2)	4,718	4,304	4,143	4,468	*8.8*	*12.2*	*5.3*
m340 (m^2)	7,802	5,790	5,481	6,107	*25.8*	*29.8*	*21.7*
m350 (m^2)	6,155	5,824	5,730	5,919	*5.4*	*6.9*	*3.8*
m360 (m^2)	1,996	2,290	2,197	2,385	*14.7*	*10.1*	*19.5*
k310 (€)	237,514	178,250	120,801	258,905	*25.0*	*49.1*	*9.0*
k320 (€)	468,382	569,881	510,121	633,868	*21.7*	*8.9*	*35.3*
k330 (€)	2,439,705	3,052,560	2,634,109	3,531,879	*25.1*	*8.0*	*44.8*
k340 (€)	1,613,306	2,002,637	1,830,356	2,185,558	*24.1*	*13.5*	*35.5*
k350 (€)	1,836,746	2,362,352	2,184,431	2,548,852	*28.6*	*18.9*	*38.8*
k360 (€)	503,369	578,663	495,664	673,805	*15.0*	*1.5*	*33.9*
k370 (€)	204,611	292,890	213,914	395,090	*43.1*	*4.5*	*93.1*
k390 (€)	255,464	434,998	344,882	544,307	*70.3*	*35.0*	*113.1*
k410 (€)	409,557	600,936	515,732	697,625	*46.7*	*25.9*	*70.3*
k420 (€)	452,667	514,458	443,645	593,774	*13.7*	*2.0*	*31.2*
k430 (€)	1,106,102	442,567	272,529	703,197	*60.0*	*75.4*	*36.4*
k440 (€)	1,216,830	922,381	737,473	1,147,098	*24.2*	*39.4*	*5.7*
k450 (€)	1,062,591	193,199	134,214	276,369	*81.8*	*87.4*	*74.0*
k460 (€)	219,244	193,516	157,137	234,884	*11.7*	*28.3*	*7.1*
k470 (€)	187,492	162,642	84,251	313,972	*13.3*	*55.1*	*67.5*
k480 (€)	118,928	196,846	123,225	287,631	*65.5*	*3.6*	*141.9*
k490 (€)	-	-	-	-	-	-	-
*k300** (€)	*7,559,098*	*9,472,230*	*8,334,278*	*10,772,264*	*25.3*	*10.3*	*42.5*
*k400** (€)	*4,773,412*	*3,226,545*	*2,468,206*	*4,254,550*	*32.4*	*48.3*	*10.9*
k300+k400** (€)	*12,332,510*	*12,698,775*	*10,802,484*	*15,026,814*	*3.0*	*12.4*	*21.8*
CD (days)	638	551	465	648	*13.6*	*27.1*	*1.6*

However, the readers and industry practitioners should bear in mind that the calculated values would be *estimated* values. Statistically, no estimation could be perfect. But with the help of advanced statistical techniques it is possible to quantify the amount of insecurity inherent in the prediction.

Therefore, when predicting a value, the interval that probably includes the true value should always be calculated. This gives *an idea* about how good the estimation is and about *the range of possible values* that actually *might result* in the end. These are the reasons motivating the consideration of confidence intervals.

A distinction has to be made between to different ways of prediction. Either an average value can be predicted. For example, the interest lies in getting information about the expected cost of external walls of projects using determined set of predictors with certain values. This is equal to predicting the average cost of external walls those projects have. The second type of prediction applies to specific single values. If an architect wants to know what value of interest he can expect for a specific project which he is planning at the moment, he can be less certain about the prediction because one single project is more affected by error related variation than the mean of a number of projects.

Accordingly, there are two ways to calculate confidence intervals, either for a mean response value or for a specific predicted value. For the practical aims of the results, it is considered to calculate prediction intervals of mean response value. For the calculation of confidence intervals an accuracy level of *5%* is selected throughout.

One can construct a confidence interval for the fitted response for a set of predictor values, $x_{01}, x_{02}, \ldots, x_{0p}$. Define the vector x_0 as

$$x_0 = \begin{pmatrix} 1 \\ x_{01} \\ x_{02} \\ \vdots \\ x_{0p} \end{pmatrix}$$

The fitted value at that point is

$$\hat{y_0} = x_0^{'} \times \hat{\beta}$$

$\hat{y_0}$is an unbiased estimator of $E(y \mid x_0)$, and the variance of $\hat{y_0}$ is

$$var(\hat{y_0}) = \sigma^2 x_0^{'}(X^{'}X)^{-1}x_0$$

Therefore, the $100(1-\alpha)\%$ confidence interval for the fitted response at

$x_{01}, x_{02}, \ldots, x_{0p}$ is

$$\hat{y_0} \pm t_{n-p-1,\alpha/2} \times \sqrt{\hat{\sigma^2} x_0^{'}(X^{'}X)^{-1}x_0} \tag{7.3}$$

Demonstration of confidence intervals would be conducted for linear regression models. Confidence intervals provided valid results in case the error terms of a model (residuals) follow a normal distribution. This was also one of the crucial assumptions of linear regression models. In order to fulfill this assumption, variable transformation in response variables were performed. Therefore, *standard error of fits* in transformed scale could be utilized to compute ranges of our estimations at 95% level.

In equation 7.3 the term $\sqrt{\hat{\sigma^2} x_0^{'}(X^{'}X)^{-1}x_0}$ refers to *standard error of the fit*, whereas $t_{n-p-1,\alpha/2}$ indicates the value of t scores under $n-p-1$ degrees of freedom and $\alpha/2$ probability.

In the case of 1300-082, external wall area prediction model fit was calculated previously as:

$$\hat{y_{m330}} = m330^{0.599167} = 150.416721$$

25 out of 600 observations were determined as outliers through modeling phase. This left 575 objects that involve in the final modeling analysis for external wall area. 7 numerical and 1 categorical variable with 5 levels were determined as the significant drivers of external wall area. This aggregated 563 degrees of freedom. Corresponding value of t score for 563 degrees of freedom and 0.025 probability can be read from relevant statistic tables as follows:

$$t_{(n-p-1),(\alpha/2)} = t_{(575-11-1),(0.05/2)} = t_{(563),(0.025)} \cong 1.9640$$

Computation of standard error of the fit in detail results as 1.73266. By using equation 7.3, 95% confidence intervals of external walls area is calculated as:

$$CI_{m330} = 150.416721 \pm 1.9640 \times 1.73266$$

$$CI_{m330,lower} = 147.024$$

$$CI_{m330,upper} = 153.830$$

Recall that response variable was transformed to fulfill assumptions of linear regression model. When back transformation was applied, lower and upper boundaries were calculated as follows:

$$CI_{m330,lower} = (147.024)^{1/0.599167} = 4143.21\text{m}^2$$

$$CI_{m330,upper} = (153.830)^{1/0.599167} = 4468.27\text{m}^2$$

For the case of 1300-082, collective results of confidence intervals for linear regression models were given in Table 7.6.

Sonmez (2008) argued that conceptual estimates should provide an interval along with point estimates for practical aims. This is due to the high level of uncertainty inherent to building works. Particularly for German practice, information flow between an architect and a client is an iterative process. It was previously discussed that rarely building projects are conducted completely parallel to the initial requirements of clients or suggestions by architects. Through conception to finalization of design, sometimes even after commence of practical construction, decisions regarding building is subject to high level of variations. This is the main reason to provide an interval for the conceptual estimates along with point predictions.

Another important reason shall be underlined. Providing an interval leaves more space for client. This is to say, when an architect is able to provide *possible* interval to a client, in which *expected* cost of structure lies between, than the decision maker, client, can behave according to its perception of risk. Considering the case of 1300-082, this interval was calculated approximately between 10.8 to 15 million € with the expected value of 12.7 million €. If the client was a *risk taker* than it could be expected that he would consider cost of structure to be between 10.8 to 12.7 million €. In case, the client is a *risk avoider* than it was likely that he would consider pessimistic scenario and expected the cost of structure somewhere between 12.7 to 15 million €. And finally in case that the client was *risk neutral* the cost of structure was likely to be considered around the expected average value of 12.7 million €. Actual value of cost of structure was observed approximately as 12.3 million €. Thereof, the result suggested a risk neutral client would benefit the results of models most for the case of 1300-082.

On the other hand, ability of considering ranges for cost of structure allows a client a great deal of financial flexibility. In consistent with conditional availability of financial sources at the given point of decision to be made, a client may take risks while sharing a certain amount of contingency. Once a project progresses and reaches to a certain level of detail where predictions can be made with more confidence, he may allow his contingency to be used in other activities or employed by the same project.

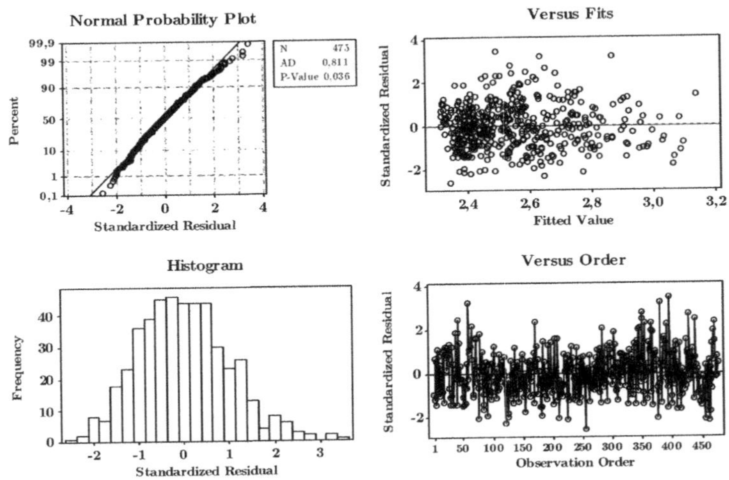

Figure 7.5: Residual plots of linear regression model for cost of external walls

For the demonstrated cases of its practical applications and benefits, we agree Sonmez (2008) that range estimates showed more practical relevance to point estimates for the conceptual stage estimates. Yet, our ideas separated in terms of applied methodology through computation of ranges. Sonmez (2008) argued *bootstrap sampling* method, relied on the main idea of generating artificial data set that shows identical properties to the original ones, was more adequate technique for range estimates. Sonmez (2008) employed generated samples to estimate parameters of regression models repeatedly for given number of observations in the main and eventually generated samples. This was to say, when one had an original data set consists of 40 observations than the same model with 40 different parameters would be estimated. Predicting the value of interests by using the same model with different parameters eventually resulted in altered fit values for a given number of generated data sets. Finally, one could compute the range values of fits based on the descriptive statistics of predicted values.

It was agreed that this was theoretically valid and adequate technique to compute range intervals. However, ease of practical implementation was dubious when the number of observations in the sample used through modeling in this case was more than 400. Considering previously exemplified cost model of external wall, Figure 7.5 illustrates that a sample of 475 observations was employed to determine the relationship best predicted cost of external walls. Please recall that variable transformation was applied to the response variable. It was evident with normal probability plot, histogram, and moreover Anderson Darling test - clear measure of normality -, that the error terms did not violate normality. In this case, assumptions of confidence intervals were fulfilled. Thereof, statistical theory could be employed instead of *bootstrap* which involved in time consuming simulation and training processes.

In conclusion, the results shown in Table 6.1 and 6.2 suggested predicting the value of interests by linear regression models. And for linear regression models, normality of error terms was fulfilled after variable transformation is applied to the response (take Figure 7.5 as an example). In case underlying assumptions were considered we argued that employment of confidence intervals over bootstrap method would be preferred due to ease of its practical implementation. The crucial message of this section was that for predictions, like the ones relevant to this research, the possible prediction error inherent in the process itself and therefore the security of the prediction had to be kept in mind. However, using statistical techniques it was a clear advantage to be able to quantify this prediction error and therefore account for it, if necessary.

8 Conclusions

Cost of structure and construction duration are two strategic determinants of building projects along with quality. Objective estimations of these determinants are crucial when establishing the basis of a project. These estimations serve as foundation for budgeting, planning, executing, monitoring and even for any litigation aims.

The research was designed, aiming to form an objective basis for estimating as well as increase prediction accuracy of project determinants, cost of structure and construction duration, to an extent for the population of interest, in this case *new building construction in Germany.* In this context, empirical investigation of historical project information was conducted. Multiple linear regression (LRM) and artificial neural networks (ANN) were adopted as the methods of empirical analysis.

German building industry is highly standardized by means of construction cost data identification, classification and storage. The standards present a hierarchical structure of construction cost data. Accordingly, *cost of structure*, corresponds to construction cost in German practice, contains 17 *second level* cost groups. *First level* cost groups, namely cost of structure - construction works and cost of structure - services, are formed by aggregation of *second level* cost groups. On the other hand, the *reference unit quantities* of second level cost groups are defined by the standards. For instance, external wall area is defined as the reference unit quantity for cost of external walls.

The regulation that describes the fee band of architects (HOAI) presents a guideline of tasks to be conducted from planning to completion of a project. The results of research undertaken is strictly limited to be used through first HOAI phase, establishing the basis of a project in which only macro project determinants, such as gross external floor area, are available to an architect to perform his early estimations.

Tightly standardized structure of German construction cost identification led us to adopt a novel approach, *multi-way* forecast. Initial investigation of the data set revealed a strong pattern between reference unit quantities and macro project determinants. This indicated that reference unit quantities could have been predicted with some confidence by using macro project determinants which were assumed to be available to an architect through establishing the basis of a project. This led to form a ground to hypothesize that, these predictions might have emerged to explain a larger portion of variance in cost groups of structure, which in turn might have influenced an increase in prediction accuracy of project determinants, to an extent. This iterative process is called *multi-way* approach. In order to test the hypothesis of whether the increase in precision of estimations is substantial, conventional *one-step ahead* predictions was conducted by using identical data sets. In one-step ahead approach, it was assumed that the architect is limited to macro projects determinants to predict cost of structure and construction duration. Four alternative scenarios were considered:

- Framework #1 adopted *multi-way* approach, aimed at predicting *second level* cost groups of structure. Total cost of structure was computed by aggregating second level cost groups of structure.

- Framework #2 adopted *multi-way* approach, aimed at predicting *first level* cost groups of structure. Total cost of structure was computed by aggregating first level cost groups of structure.

- Framework #3 adopted *one-step ahead* approach, aimed at predicting *first level* cost groups of structure. Total cost of structure was calculated by aggregating first level cost groups of structure.

- Framework #4 adopted *one-step* ahead approach, aimed at predicting cost of structure, directly.

Identical flow of information and method of analysis was also used in prediction of construction duration. Two different frameworks were constructed as follows:

- Framework #5, adopted *multi-way* approach. It could be regarded as an extension to Framework #1 where fitted values of second level cost groups were entered as additional system inputs to predict construction duration.

- Framework #6, adopted *one-step ahead* approach, aimed at predicting construction duration with limited information available.

The training sample, consisted of 600 objects, and the test sample, contained 57 projects, were used for modeling and validation aims, respectively.

In the first part of this thesis, linear regression and neural network models of reference unit quantities, second and first level cost groups of structure, and construction duration were developed. It is important to emphasize that all types of the models were developed by using exactly the same number of projects, same data sets and same procedures.

For the first part, linear regression was implemented to obtain parsimonious models. Stepwise regression procedure was adopted to select the variables which were retained in the initial model at 10% significance level. Later, categorical variables were included in the model. Redundant variables, that were not contributed to the models at 5% significance level, were excluded from the final models. The parameters of final model were estimated by ordinary least square estimation. Prediction accuracies of linear models were validated on the test sample. Individual results were discussed.

Later, ANN models were developed by using significant factors of linear regression models as input nodes. Type and architecture of a network model were determined based on selecting from different alternatives which reached to the lowest prediction error in the test sample for the value of interest. Individual comparisons of prediction performance indicators between linear regression and neural network models were conducted.

Main findings of the first part can be outlined as follows:

- Gross external floor area (GEFA), average floor size (AFS) and facility type (FAC) are the significant factors which were retained in all reference unit quantity models.

- Linear regression and neural network models did not show significant differences of prediction accuracy for excavation volume (m310) and internal walls area (m350), when mean absolute percentage error (MAPE) is considered as the measure of accuracy. However, the findings suggested, neural network models provided superior prediction accuracy in comparison to linear regression models for the rest of reference unit quantities, namely foundation area (m320), external walls area (m330), floors and ceilings area (m350) and roof area (m360).

- In the first step of the analysis, estimated values of reference unit quantities were gained. Results by predictive models of *second level* cost groups suggested these quantities made significant contribution to account for substantial amount of the variance in the responses, especially for cost groups of construction works.

- Adopting *multi-way approach*, linear regression models reached to smaller prediction errors than neural network models in majority of the *second level* cost groups of structure. The number of outliers detected in the test sample did not show significant differences. Dispersion in mean absolute percentage errors of second level cost groups was significantly lower than its counterpart, neural network models. The results also revealed that observed prediction error in cost groups of construction works was significantly lower than cost groups of services. It can be argued predicted reference unit quantities, which later fed back system inputs in accordance with the principle of multi-way approach, were more relevant to construction works rather than services.

- Results by predictive models of *first level* cost groups of structure suggested, gross building volume (GBV), developed plot area (DPA), facility type (FAC) and client type (CT) were the significant drivers, appeared in all three models, when *multi-way approach* was considered. Besides, gross external floor area (GEFA), average floor size (AFS), *fitted* values of external ($m330_{fit}$) and internal walls area ($m340_{fit}$) were retained in, at least, two out of three models.

- Using *multi-way approach*, results showed linear regression models provided superior prediction accuracy to neural network models as well as presumed decrease in dispersion of the error terms in the test sample, when the aim was set to estimate *first level* cost groups of structure.

- Adopting *one-step ahead* approach, gross external floor area (GEFA), plot area (PA), average floor size (AFS), facility type (FAC) and client type (CT) were the significant drivers of *first level* cost groups of structure. Number of storeys (NoS) and market conditions (MC) appeared at least two out of three models.

- Results suggested pattern recognition capacity of linear regression models was superior to neural network models, for the values of interest, in this case *first level* cost groups of structure, when intention was set to employ *one-step ahead* approach. This was due smaller number of outliers detected and lower error observed through predicting the values of interest in unseen data, the test sample.

- Regardless of the prediction approach, client type (CT), facility type (FAC), construction start date in years (CSD), and number of storey blow ground (NoS_{bg}) were determined as the significant drivers of construction duration. In one step ahead prediction, the model also included gross external floor area (GEFA). The variable became redundant as *fitted* values of second level cost groups along with gross building volume (GBV) were retained in the final model when multi-way approach was considered. It can be argued that explanatory power of gross external floor area was isolated by those of variables retained in the multi-way model. Above all, including construction start date as a predictor, both models suggested time dependency of construction duration.

- Considering mean absolute percentage error (MAPE) as the measure of prediction accuracy, the difference between alternative method of analysis was insignificant. In majority of the reference unit quantities predictions neural network models provided superior results. When cost estimation was considered, linear regression models were favored. Through duration estimation a slight difference between two methods were present. Therefore it can be concluded that error rates in prediction of project determinants were reasoned by neither functional form of pre assumed relationships in between variables, nor model assumptions to be fulfilled. The reason was, presumably, lacking of crucial information which might account for the additional variance in the value of interests that could not already be covered by the information available. However, when absolute percentage error of individual test objects were investigated, dispersion of error terms was significantly higher in the case of neural network models. Thereof, it can be concluded it is a clear advantage of employing linear regression models when the aim is set for prediction.

In the second part of the study, formal test of hypothesis was conducted. Therefore, the second part investigated whether the primary aim of reducing error of predictions performed through establishing the basis of a project could have been accomplished, to an extent.

For the second part, individual model results were aggregated in accordance with offered frameworks of cost and duration estimation with respect to alternative methods of analysis. A formal test of hypothesis was conducted to reveal whether differences among MAPEs of the test sample is statistically significant at 5% level. This was to investigate if the outcomes of this part can be generalized to overall population or the role of chance can not be ignored. ANalysis Of VAriance (ANOVA) was selected as the appropriate statistical procedure to run the formal test of hypothesis.

Main findings of the second part can be outlined as follows:

- Aiming cost estimation, Framework #2 reached to the lowest prediction error in the test sample among others when linear regression method was adopted. In addition, employing linear regression as the method of analysis, frameworks that adopted *multi-way* approach provided higher prediction accuracies when compared to those adopted one-step ahead approach.

- Considering linear regression models, the difference in prediction accuracy between Framework #2 and #1 was not significant. It was argued, it is a clear advantage for practical implementations to be able to provide not only reference unit quantities but also second level cost groups to the architect. Thereof, Framework #1 shall be selected when the aim is set for cost estimation.

- The results of ANOVA concluded that there are *no* differences among the mean absolute percentage errors of construction cost frameworks over test sample at the 5% level of significance. The null hypothesis can not be rejected. This was basically influenced by insufficient size of the test sample. Accordingly, to detect the significance of differences at these margins and to generalize the results for overall population, 300 and more objects shall be considered for testing aims.

- Aiming duration estimation, Framework #6 reached to the lowest prediction error in the test sample among others when artificial neural network method was adopted. However, the shortcomings of this method were previously discussed. Therefore, Framework #5 using linear regression was entitled when the aim is set for duration estimation.

- It is evident by the findings that multi-way approach influenced an increase in prediction accuracy when compared to one-step ahead approach. Yet, this difference was denied as significant by the formal test of hypothesis. Similar to the issue reported for cost estimation, insufficient size of the sample plays an influential role.

- It can be argued using multi-way approach, primary aim of reducing error of predictions performed at the conceptual design phase was accomplished, to an extent. However, differences were not substantial, statistically. Yet, for practical implementations, even slight improvements are very valuable in building construction.

- Ruf (2003) indicated prediction errors of the conceptual estimates can be as high as $\pm 30\%$ at the initial phases of developments. Therefore, it can be argued that the results of this study presents practically applicable results. Considering cost estimation, selected framework offer 27% improvement to the threshold value by Ruf (2003). When the aim is set for duration estimation, the mean error reached to the borderline.

In the third part of the study, practical implementation of the results were demonstrated on a randomly selected case. What is more to point, computation and interpretation of confidence intervals were presented. The crucial message of third part was that for predictions, like the ones relevant to this research, the possible prediction error inherent in the process itself and therefore the security of the prediction had to be kept in mind. However, using statistical techniques it was a clear advantage to be able to quantify this prediction error and therefore account for it, if necessary.

Last but not least, further research may focus on implementing alternative methods of analysis such as case-based reasoning for the problem of interest. Another idea may be to develop an index defines structural quality of buildings, which in turn may be used to adjust the estimation through establishing the basis of a project. It would also be a valuable contribution to study regarding the average departures of cost of structure, construction duration between planned values through establishing the basis of a project and actual values incurred in the construction site for German practice.

References

Adeli, H., & Wu, M. 1998. Regularization neural network for construction cost estimation. *Journal of Construction Engineering and Management*, **124**(1), 18–24.

Bhokha, S., & Ogunlana, S.O. 1999. Application of artificial neural network to forecast construction duration of buildings at the predesign stage. *Engineering, Construction and Architectural Management*, **6**(2), 133–144.

Bordoli, D.W., & Baldwin, A.N. 1998. A methodology for assessing construction project delays. *Construction Management & Economics*, **16**(3), 327–337.

Bowen, PA, & Edwards, PJ. 1985. Cost modelling and price forecasting: practice and theory in perspective. *Construction Management and Economics*, **3**(3), 199–215.

Bromilow, FJ. 1969. Contract time performance expectations and the reality. *Pages 70–80 of: Building Forum*, vol. 1.

Bromilow, FJ. 1974. Measurement and scheduling of construction time and cost performance in the building industry. *The Chartered Builder*, **10**(9), 57–65.

Chan, A.P.C. 1999. Modelling building durations in Hong Kong. *Construction Management and Economics*, **17**(2), 189–196.

Chan, A.P.C. 2001. Time-cost relationship of public sector projects in Malaysia. *International Journal of Project Management*, **19**(4), 223–229.

Chan, D.W.M., & Kumaraswamy, M.M. 1995. A study of the factors affecting construction durations in Hong Kong. *Construction management and Economics*, **13**(4), 319–333.

Chan, D.W.M., & Kumaraswamy, M.M. 1999. Modelling and predicting construction durations in Hong Kong public housing. *Construction Management and Economics*, **17**(3), 351–362.

Chatterjee, S., & Hadi, A.S. 2006. *Regression analysis by example*. John Wiley and Sons.

Cox, A. 1996. Relational competence and strategic procurement management: Towards an entrepreneurial and contractual theory of the firm. *European Journal of Purchasing & Supply Management*, **2**(1), 57–70.

Cox, A., & Thompson, I. 1997. Fit for purpose'contractual relations: determining a theoretical framework for construction projects. *European journal of purchasing & supply management*, **3**(3), 127–135.

Cox, Reavis, & Goodman, Charles S. 1956. Marketing of Housebuilding Materials. *Journal of Marketing*, **21**(1), pp. 36–61.

Cybenko, G. 1989. Approximation by superpositions of a sigmoidal function. *Mathematics of Control, Signals, and Systems (MCSS)*, **2**(4), 303–314.

DESTATIS. 2012 (November). *Statistische Aemter des Bundes und des Laender [Statistical offices of the Federation and the states]*. Online.

DIN. 2005. *DIN 277-3 Grundflächen und Rauminhalte von Bauwerken im Hochbau [Plan areas and volumes in building]*.

DIN. 2008. *DIN 276-1 Kosten im Hochbau [Building costs - Part1: Building construction]*.

Dubois, Anna, & Gadde, Lars-Erik. 2002. The construction industry as a loosely coupled system: implications for productivity and innovation. *Construction Management & Economics*, **20**(7), 621–631.

Duncan, W.R. 1996. *A guide to the project management body of knowledge*. Project Management Institute.

Dursun, O. 2008. *Time-cost relationship of residential building projects in Istanbul*. M.Phil. thesis, The University of Reading.

Dursun, O. 2009. *Modelling construction duration for building projects in Istanbul*. M.Phil. thesis, Istanbul Technical University.

Dursun, Onur, & Stoy, Christian. 2011. Time-cost relationship of building projects: statistical adequacy of categorization with respect to project location. *Construction Management & Economics*, **29**(1), 97 – 106.

Dursun, Onur, & Stoy, Christian. 2012. Determinants of construction duration for building projects in Germany. *Engineering, Construction and Architectural Management*, **19**, 444–468.

Emsley, M.W., Lowe, D.J., Duff, A.R., Harding, A., & Hickson, A. 2002. Data modelling and the application of a neural network approach to the prediction of total construction costs. *Construction Management & Economics*, **20**(6), 465–472.

Fellows, R., & Liu, A. 2008. *Research methods for construction*. Blackwell Pub.

FIEC. 2011. *Construction in Europe - key figures*.

Finney, S.W. 2001. *Automated method for a takeoff estimate of construction drawings*.

Fortune, C., & Lees, M. 1994. Early cost advice for clients - the practitioner's verdict. *Pages 422–432 of: Tenth Annual Conference, ARCOM, Loughborough*.

Gann, D.M. 1996. Construction as a manufacturing process? Similarities and differences between industrialized housing and car production in Japan. *Construction Management & Economics*, **14**(5), 437–450.

Gidado, K. I. 1996. Project complexity: The focal point of construction production planning. *Construction Management & Economics*, **14**(3), 213 – 225.

Glenigan. 2011. UK Construction Industry KPIs - Industry Performance Report.

Hegazy, T., Fazio, P., & Moselhi, O. 1994. Developing Practical Neural Network Applications Using Back-Propagation. *Computer-Aided Civil and Infrastructure Engineering*, **9**(2), 145–159.

Hegazy, Tarek, & Ayed, Amr. 1998. Neural network model for parametric cost estimation of highway projects. *Journal of Construction Engineeing and Management*, **124**, 210–218.

Hill, Tim, Marquez, Leorey, O'Connor, Marcus, & Remus, William. 1994. Artificial neural network models for forecasting and decision making. *International Journal of Forecasting*, **10**(1), 5 – 15.

Hoffman, G.J., Thal Jr, A.E., Webb, T.S., & Weir, J.D. 2007. Estimating performance time for construction projects. *Journal of Management in Engineering*, **23**, 193.

Hornik, K., Stinchcombe, M., & White, H. 1989. Multilayer feedforward networks are universal approximators. *Neural networks*, **2**(5), 359–366.

Ireland, V. 1985. The role of managerial actions in the cost, time and quality performance of high-rise commercial building projects. *Construction Management and Economics*, **3**(1), 59–87.

Irie, B., & Miyake, S. 1988. Capabilities of three-layered perceptrons. *Pages 641–648 of: Neural Networks, 1988., IEEE International Conference on.* IEEE.

Kaka, A., & Price, ADF. 1991. Relationship between value and duration of construction projects. *Construction Management and Economics*, **9**(4), 383–400.

Kang, S.Y. 1991. *An investigation of the use of feedforward neural networks for forecasting.* Ph.D. thesis, Kent State University.

Karanci, H. 2010. *A Comparative Study of Regression Analysis, Neural Networks and Case-Based Reasoning for Early Range Cost Estimation of Mass Housing Projects.* M.Phil. thesis, Middle East Technical University.

Khosrowshahi, F., & Kaka, A.P. 1996. Estimation of project total cost and duration for housing projects in the UK. *Building and Environment*, **31**(4), 375–383.

Kim, G.H., An, S.H., & Kang, K.I. 2004. Comparison of construction cost estimating models based on regression analysis, neural networks, and case-based reasoning. *Building and environment*, **39**(10), 1235–1242.

Kouskoulas, V., & Koehn, E. 1974. Predesign cost-estimation function for buildings. *Journal of the Construction Division*, **100**(4), 589–604.

Kumaraswamy, M.M., & Chan, D.W.M. 1995. Determinants of construction duration. *Construction Management and Economics*, **13**(3), 209–217.

Lapedes, A., & Farber, R. 1988. How neural nets work. *Pages 442–456 of: Neural information processing systems.*

Li, H. 1995. Neural networks for construction cost estimation. *Building research and information*, **23**(5), 279–284.

Li, H., Shen, QP, & Love, P.E.D. 2005. Cost modelling of office buildings in Hong Kong: an exploratory study. *Facilities*, **23**(9/10), 438–452.

Love, P.E.D., Tse, R.Y.C., & Edwards, D.J. 2005. Time-Cost Relationships in Australian Building Construction Projects. *Journal of Construction Engineering and Management*, **131**, 187.

Low, P.S., & Mok, S.H. 1999. The application of JIT philosophy to construction: a case study in site layout. *Construction Management & Economics*, **17**(5), 657–668.

Lowe, D.J., Emsley, M.W., & Harding, A. 2006. Predicting construction cost using multiple regression techniques. *Journal of Construction Engineering and Management*, **132**, 750.

Minitab, Inc. 2010. *MINITAB 16 reference manual.* Minitab.

Murat Gunaydin, H., & Zeynep Dogan, S. 2004. A neural network approach for early cost estimation of structural systems of buildings. *International Journal of Project Management*, **22**(7), 595–602.

Newton, S. 1991. An agenda for cost modelling research. *Construction Management and Economics*, **9**(2), 97–112.

Ng, S.T., Mak, M.M.Y., Skitmore, R.M., Lam, K.C., & Varnam, M. 2001. The predictive ability of Bromilow's time-cost model. *Construction Management and Economics*, **19**(2), 165–173.

Ogunsemi, DR, & Jagboro, GO. 2006. Time-cost model for building projects in Nigeria. *Construction Management and Economics*, **24**(3), 253–258.

Park, J., & Sandberg, I.W. 1991. Universal approximation using radial-basis-function networks. *Neural computation*, **3**(2), 246–257.

Ross, A. 1998. *A price on design.* London: The Royal Institution of Chartered Surveyors.

Ruf, HU. 2003. Verfahren der Kostenermittlung, Kostenkontrolle und Kostensteuerung. *BKI Handbuch Kostenplanung im Hochbau, Stuttgart: BKI*, **2**, 90–109.

Sato, F., Shimada, Y., Selaru, F.M., Shibata, D., Maeda, M., Watanabe, G., Mori, Y., Stass, S.A., Imamura, M., & Meltzer, S.J. 2005. Prediction of survival in patients with esophageal carcinoma using artificial neural networks. *Cancer*, **103**(8), 1596–1605.

Shammas-Toma, M., Seymour, D., & Clark, L. 1998. Obstacles to implementing total quality management in the UK construction industry. *Construction Management & Economics*, **16**(2), 177–192.

Sharda, R. 1994. Neural networks for the MS/OR analyst: An application bibliography. *Interfaces*, **24**(2), 116–130.

Shastri, V., Rabelo, L.C., Onjeyekwe, E., & Vila, J. 1998. Device-independent Color Correction for Multimedia Applications Using Neural Networks and Abductive Modeling Approaches. *Expert Systems*, **15**(2), 110–119.

Shin, Y., & Ghosh, J. 1995. Ridge polynomial networks. *Neural Networks, IEEE Transactions on*, **6**(3), 610–622.

Shirazi, B., Langford, DA, & Rowlinson, SM. 1996. Organizational structures in the construction industry. *Construction Management & Economics*, **14**(3), 199–212.

Skitmore, M. R., & Ng, T. S. 2003. Forecast models for actual construction time and cost. *Building and environment*, **38**(8), 1075–1083.

Sonmez, R. 2004. Conceptual cost estimation of building projects with regression analysis and neural networks. *Canadian Journal of Civil Engineering*, **31**(4), 677–683.

Sonmez, R. 2011. Range estimation of construction costs using neural networks with bootstrap prediction intervals. *Expert Systems with Applications*, **38**, 9913–9917.

Sonmez, Rifat. 2008. Parametric Range Estimating of Building Costs Using Regression Models and Bootstrap. *Journal of Construction Engineering & Management*, **134**(12), 1011 – 1016.

Specht, Donald F. 1991. A general regression neural network. *Neural Networks, IEEE Transactions on*, **2**(6), 568–576.

Stoy, C. 2007. *Building costs planning at the early project phases [Baukostenplanung in fruehen Projektphasen]*. vdf Hochschulverlag AG.

Stoy, C., Dreier, F., & Schalcher, H.R. 2007. Construction duration of residential building projects in Germany. *Engineering, Construction and Architectural Management*, **14**(1), 52–64.

Stoy, C., Pollalis, S., & Dursun, O. 2012. A concept for developing construction element cost models for German residential building projects. *International Journal of Project Organisation and Management*, **4**(1), 38–53.

Stoy, Christian, Pollalis, Spiro, & Schalcher, Hans-Rudolf. 2008. Drivers for Cost Estimating in Early Design: Case Study of Residential Construction. *Journal of Construction Engineering & Management*, **134**(1), 32 – 39.

Tabachnick, B.G., Fidell, L.S., & Osterlind, S.J. 2001. *Using multivariate statistics*. Allyn and Bacon Boston.

Tam, CM, & Fang, C.F. 1999. Comparative cost analysis of using high-performance concrete in tall building construction by artificial neural networks. *ACI Structural Journal*, **96**(6).

Vrijhoef, R., & Koskela, L. 2000. The four roles of supply chain management in construction. *European Journal of Purchasing & Supply Management*, **6**(3-4), 169–178.

Walker, D.H.T. 1995. An investigation into construction time performance. *Construction Management and Economics*, **13**(3), 263–274.

Weigend, A.S., Huberman, B.A., & Rumelhart, D.E. 1992. Predicting sunspots and exchange rates with connectionist networks. *Pages 395–395 of: Santa Fe Institute Studies in the Sciences of Complexity*, vol. 12. ADDISON-WESLEY PUBLISHING CO.

Widrow, B., Rumelhart, D.E., & Lehr, M.A. 1994. Neural networks: Applications in industry, business and science. *Communications of the ACM*, **37**(3), 93–105.

Winch, G. 1987. The construction firm and the construction process: the allocation of resources to the construction project. *Managing construction worldwide*, **2**.

Wong, F.S. 1991. Time series forecasting using backpropagation neural networks. *Neurocomputing*, **2**(4), 147–159.

Yeong, C. M. 1994. *Time and cost performance of building contracts in Australia and Malaysia.* M.Phil. thesis, The University of South Australia.

Zhang, G., Eddy Patuwo, B., & Y Hu, M. 1998. Forecasting with artificial neural networks: The state of the art. *International journal of forecasting*, **14**(1), 35–62.

Zhang, Q., & Benveniste, A. 1992. Wavelet networks. *Neural Networks, IEEE Transactions on*, **3**(6), 889–898.

Zhang, X. 1994. Time series analysis and prediction by neural networks. *Optimization Methods and Software*, **4**(2), 151–170.

Printed and bound by CPI Group (UK) Ltd, Croydon, CR0 4YY
13/09/2026
14961093-0006